Missing Data Analysis in Practice

CHAPMAN & HALL/CRC
Interdisciplinary Statistics Series

Series editors: N. Keiding, B.J.T. Morgan, C.K. Wikle, P. van der Heijden

Published titles

Published titles

Chapman & Hall/CRC
Interdisciplinary Statistics Series

Missing Data Analysis in Practice

Trivellore Raghunathan

University of Michigan
Ann Arbor, Michigan, USA

CRC Press
Taylor & Francis Group
Boca Raton   London   New York

CRC Press is an imprint of the
Taylor & Francis Group, an **informa** business

A CHAPMAN & HALL BOOK

CRC Press
Taylor & Francis Group
6000 Broken Sound Parkway NW, Suite 300
Boca Raton, FL 33487-2742

© 2016 by Taylor & Francis Group, LLC
CRC Press is an imprint of Taylor & Francis Group, an Informa business

No claim to original U.S. Government works

Printed on acid-free paper
Version Date: 20150915

International Standard Book Number-13: 978-1-4822-1192-4 (Hardback)

Visit the Taylor & Francis Web site at
http://www.taylorandfrancis.com

and the CRC Press Web site at
http://www.crcpress.com

To my Teachers
and my Students.

Contents

List of Tables

List of Figures

Preface

Missing data problems are ubiquitous and as old as data analysis itself. Researchers have been dealing with missing data in various ways and in a somewhat *ad hoc* manner. While some *ad hoc* procedures have taken roots, principled approaches such as maximum likelihood were developed in the middle of the last century. Lack of computational power and to some extent sheer inertia against any change led to a confusing landscape. Many principled methods were not implemented in familiar software packages. The last quarter of the last century witnessed massive changes in the computational landscape and large scale applications of the statistical methodology. This allowed methodological development for handling incomplete data, Bayesian methods, replication or resampling methods, complex modeling and their implementation to a desktop, to a laptop and even to a palm of one hand!

Now researchers are like children in a toy store! Too many methods, too many implementations and still a confusing landscape. Every missing data method makes assumptions. This is not bad and is just inherent in making inferences with incomplete data. The design of a study may be under the control of the investigator, but responses from subjects in the study or their cooperation to provide responses are not. In fact, it can be argued that assumptions are inherent in any statistical inferential activity as a projection from a sample to the population or the phenomenon. It is important, therefore, to know those assumptions and to discriminatively assess the suitability of assumptions in the specific context.

This book describes several easy-to-implement approaches, discusses the underlying assumptions, provides some practical means for assessing these assumptions and then suggests implementation. Numerous approximations are used as no practical problem comes in a nice theoretically elegant mold. The theory is described using heuristics rather than rigor. Actual and simulated data sets are used to illustrate important concepts.

This book uses ideas from both Frequentist and Bayesian perspectives but has a definite Bayesian flavor. A statistical practitioner is like a "handyman," a person interested in solving the practical problems and no good tool, frequentist or Bayesian, is wasteful. All distributional assumptions are stated so that they can be interpreted from both perspectives. The danger with this attitude is that both camps may be unhappy with the book!

Many books on incomplete data have been published over the last few years. Why another book? Some books are a bit too theoretical and suitable for students in statistics or biostatistics programs. Some books are geared towards one approach or the other and/or one software or the other. Some are too general and lack specifics on how to apply. Over the last 25 years, I have taught several one-day to two semester long courses for a variety of audiences with a range of quantitative backgrounds. I have attempted to write this book for that kind of audience as if they are all in my lecture. I have described the methods in simple terms and in technical terms. This book attempts to provide heuristic reasoning of the assumptions, theoretical understanding and practical implementation. This book represents a collective experience of research, teaching and consulting.

A two semester course in statistics including regression analysis should suffice to understand most of the material. A course in Bayesian analysis is a big plus for a practitioner as it can expand the knowledge base for tackling many practical problems. A healthy Frequentist and Bayes concoction are a great way to start the statistical practice.

I have co-taught several two-day short courses with Rod Little and have developed Power-Point slides. I have greatly benefited by this collaboration. The first four chapters of the book mostly follow the presentation in the slides. The maximum likelihood (ML) approach is discussed in brief. Of course, Little and Rubin (2002) is the best book for ML! At Michigan, Rod has been a wonderful mentor, a dear friend and a superb colleague.

Though I received solid training in mathematical statistics at the Institute of Science, Nagpur in India, and a thorough knowledge of applications at Miami University in Oxford, Ohio, my statistical being was shaped at Harvard University by Don Rubin and Art Dempster. Fred Mosteller and John Tukey (who provided great inspiration and emphasized the importance of "listening to data" during his regular summer visits to Harvard) greatly influenced me. Being a teaching assistant for Fred Mosteller provided the unique perspective on teaching, nay, learning through and with the students. To this mix add some of the smartest people, Nat Schenker, Xiao-li Meng, Andy Gelman, Alan Zaslavsky, Emery Brown, Joe Schafer, Tom Belin and many others as colleagues at the graduate school, academic brothers and the continuing relationship beyond (Am I describing an academic heaven?). I owe my deep gratitude to these and many more. A special thanks to Nat Schenker who has been a great collaborator.

I thank Dawn Reed for compiling the references and other help with manuscript preparations. I am thankful to all the students in my classes who over the years brought their missing data problems to my attention, helped me understand the practical issues and shaped this book through their comments and through challenging some assertions. These discussions led to several simulation studies for the entire class. Many such simulation studies are discussed in the book.

The first seven chapters deal with the traditional missing data problem. However, many statistical problems with no missing values can be addressed using the missing data framework. Some of these applications are discussed in Chapter 8 and thus extending the application of methods discussed in this book. Missing data methods development and its adaptation to practical problems is, therefore, a great area of research and application. Hope that material presented in the book is useful!

Trivellore Raghunathan ("Raghu")
Ann Arbor, Michigan

1

Basic Concepts

1.1 Introduction

Data collection and analysis forms the backbone of all empirical research and almost every data analysis involves variables with some missing values (which will be defined later). The missing values may arise due to unit nonresponse where a sampled subject refuses to provide any values for the variables of interest, or due to item nonresponse, where a sampled subject provides information only for some variables.

The complete-case or available-subjects is the common approach that restricts the analysis to subjects without missing values in the relevant variables. This approach, though convenient, can result in biased estimates of the parameters or population quantities because the included and excluded subjects from the analysis may differ systematically. Even if the included subjects are a random subset of the sampled subjects, the sampling error increases due to the reduced sample size.

Many *ad hoc* and naive methods are also used in practice. For example, in a multiple regression analysis with missing values in one categorical covariate, subjects with missing values are treated as a separate category when creating the dummy variables. This analysis uses all of the subjects but may seriously bias the regression coefficients for the other covariates. Another naive method involves substituting a fixed value, such as the mean, median or the mode based on respondents for all subjects, with missing values. This strategy creates artificial "peaks" in the imputed (or the completed) data set, resulting in bias in any analysis involving statistics measuring spread or dispersion, such as regression analysis.

Despite several studies demonstrating bias through theoretical and simulation investigations in using complete-case and *ad hoc* methods, they continue to be used. There may be special circumstances or assumptions under which some of these approaches may be valid, but those are hard to verify. Why bother with these approaches when better approaches are available to analyze incomplete data?

There is no assumption free method for analyzing incomplete data. Since the analysis involves something that is unknown (missing values) and something that is known (observed values), inference is always going to be conditional on the relationship between what is known and what is unknown. This relationship is often expressed through a missing data mechanism, a probability model or process that leads to missing values and its relationship to the study or analysis variables. It is very important to understand the missing data mechanisms to judge the appropriateness of an analysis procedure using the observed data.

1.2   Definition of Missing Values

A clear definition of a missing value should be established before proceeding further. Operationally (or heuristically), a value is defined to be missing for a variable if a meaningful value for the specific analysis to be performed is hidden. Examples of such situations include variables such as income, blood pressure, education, age, etc.

The definition also depends on the scientific question being answered. Consider an example where the definition of a missing value is not clear cut. Suppose that a survey is being conducted prior to an election where candidates from the major parties (A and B) are contesting for a seat. A question (X) is asked, "Are you going to vote for the party A or B?" and the response options are (1) A, (2) B and (3) Don't know. One may plan an analysis with X as having three response categories. For the projection of a winner, however, "Don't know" responses may be treated as missing values and a mechanism may be needed to classify "Don't knows" into vote for A or B.

The problem can be more complex when the following question is asked, (Y), "Are you planning to vote in the upcoming election?" with the response options (1) Yes, (2) No and (3) Don't know. Again, for some analysis the three-response category variables may be legitimate analytical variables. For the projection purposes, both variables (X and Y) have to be used and the "Don't know" response (3) in both variables have to be treated as missing values. The resolution of missing values in Y, determines the relevant population for handling the missing values in X.

Consider a longitudinal study measuring blood pressure repeatedly and some subjects dropped out of the study. If the subject is known to be alive, then the missing blood pressure measurement is a meaningful value for the analysis. If the subject is known to have died, then the blood pressure measurement is, generally, not a meaningful value for the analysis (dealing with selection due to death in the analysis is a separate issue). Even here, it is possible that some analysis may involve consideration of values such as "Had this person been alive at the time of measurement what would have been his/her blood pressure"? Such questions may arise in the competing risk analysis of two or more diseases.

An intuitive way to define the missing values for a variable in a specific analysis is to consider whether or not one should impute those values for subjects. In general, it is a good idea to flag all the imputed values to provide a flexibility in the analysis and for diagnostic purposes.

1.3   Missing Data Pattern

A pattern of missing data describes the location of the missing values in a potential complete data matrix (that is, data matrix with 100% response rate). For simplicity, consider a rectangular data matrix with rows representing subjects and columns representing variables. The rows and columns in the data matrix can be sorted or rearranged to get special patterns of missing data. The following figure illustrates various patterns of missing data.

Pattern (a) shows a monotone pattern of missing data where the variable $j = 2, 3, \ldots, p$ is observed on a subset of subjects with variable $j-1$ observed. This pattern of missing data typically arises in a longitudinal or a panel study when the drop outs from each wave are not followed in the future (generally, a bad idea).

Another common pattern is shown in Pattern (b) where the data is missing only on one variable in the analysis. Pattern (c) occurs when two files are appended where File 1 provides data on Y_1 and Y_2, and File 2 provides data

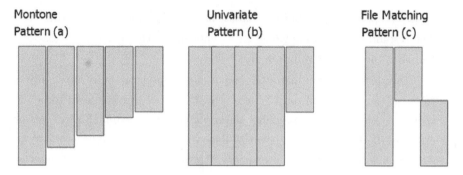

Figure 1.1: Patterns of missing data

on Y_1 and Y_3. This type of pattern also occurs in causal inference where Y_2 and Y_3 are the potential outcomes under two treatments $Y_1 = 1$ or $Y_1 = 0$, where Y_2 is not observed on those receiving treatment $Y_1 = 0$, and Y_3 is not observed on subjects receiving treatment $Y_1 = 1$.

The pattern of missing data could be exploited in the model specification or by breaking the estimation problem into simpler modular tasks. The second use of pattern is to understand the limitation of the data or identify parameters that cannot be estimated. For example, in Pattern (c), there is no information to estimate the partial correlation between Y_2 and Y_3 conditional on Y_1.

In most, if not all, practical situations, the pattern of missing data will be arbitrary or a general pattern of missing data. The methods described in this book are geared towards the general pattern of missing data but could be applied to other patterns of missing data. Whenever possible, alternative methods for a specific pattern of missing data will be suggested.

1.4   Missing Data Mechanism

To understand the concept of missing data mechanism, consider a case with single variable U with some missing values and a set of covariates V with no missing values. Let R be a response indicator taking the value 1 if U is observed and 0 if U is missing. The missing data mechanism is an assumed probabilistic or regression relationship between R and (U, V). One may view R as a "treatment assignment" in an experimental design context. This analogy will be helpful in understanding the terminology used in the missing data mechanism.

Note that the substantive data (U, V) is not fully observed because U is not fully observed. The substantive data can be decomposed into the observed data, (U_{obs}, V), consisting of the observed set of values U_{obs} on subjects who provided them, and missing data U_{mis}, consisting of unknown values of U on subjects who did not provide information about U. Since R is a binary variable, the relationship between R and $D = (U, V)$ can be expressed through the probability specification, $Pr(R = 1 | U, V)$ (the probability of observing U or the response propensity).

The values, U_{mis} are considered to be missing completely at random (MCAR), if the response propensity is a constant number across all the subjects. That is, $Pr(R = 1 | U, V) = $ constant. In the experimental design context, this mechanism is similar to the treatment assignment in a completely randomized design (CRD) where every individual in the sample has the same chance of being assigned the treatment $R = 1$.

The MCAR assumption implies that the distribution of the outcome variable is the same for both groups ($R = 0$ and $R = 1$). Thus, under MCAR the subsample corresponding to the complete-case data (U_{obs}, V_{obs}) is a simple random sample or a random subset of the full sample. There is no loss of "representativeness" of the sample for the population by restricting our analysis (U_{obs}, V_{obs}).

Generally, the complete-case analysis is valid under the MCAR mechanism. There are some exceptions. For example, the complete-case analysis yields unbiased estimates in a regression analysis with U as the dependent variable and some or all of V as independent variable even when the data are not MCAR (see Little and Rubin (2002) for more details). It is difficult to establish general conditions for the validity of complete-case analysis. Besides, the complete-case analysis usually will have larger sampling errors due to smaller sample size even if the point estimates are unbiased. Better methods (and software to implement them) are available.

The MCAR is a strong assumption. Often several factors or characteristics of subjects may influence the decision to answer survey questions. A weaker assumption is called missing at random (MAR). In the experimental design context, this mechanism may be viewed as randomized block design. Referring to the same setup with U and V, suppose that there are m individuals with the same value of $V = v$, forming a block based on V. Let m_1 and m_o be the number of respondents and nonrespondents, respectively, in this block. Under the MAR assumption, the distribution of U is the same for the m_1 respondents and the m_o nonrespondents. That is, within each block, the assignment of the label of respondent/nonrespondent is completely at random. More generally, the response propensity under MAR is $Pr(R = 1|U, V) = Pr(R = 1|U_{obs}, V)$.

Whether MAR is a reasonable assumption depends upon the correlation between U and the blocking variable V. Strong correlation between V and U implies weak MAR assumption. For example, if U is income and V consists of only age and gender, then the MAR assumption conditional on age and gender is stronger in comparison to when V consists of age, gender, race, education, occupation, employment, etc. Hence, it is critical to obtain information on correlates of variables with missing values on the full sample to match the respondents and nonrespondents as much as possible to make residual differences between U_{obs} and U_{mis} within any block minimal and random.

Ignorable missing data mechanism is defined as MCAR or MAR and also "distinctness" of the parameters in the distributions of the substantive variables (U, V) and the response indicator R given (U_{obs}, V). Suppose that $Pr(U|V, \theta)$ or $Pr(U, V|\theta)$ is the statistical model (conditional or joint distribution) that the analyst will be using if he/she had the complete data, and θ is the unknown parameter to be inferred. Suppose that $Pr(R = 1|U_{obs}, V, \phi)$ is the response propensity with unknown parameter ϕ. If θ and ϕ are not functionally related to each other (or that the knowledge of one does not provide any information about the other) then the parameters are called "distinct," and the missing data mechanism is ignorable.

For a technical description, let Θ and Φ be the parameter space for θ and ϕ, respectively. A MAR or MCAR missing data mechanism is ignorable if the joint parameter space for (θ, ϕ) is $\Theta \times \Phi$ or in the Bayesian framework, the parameters θ and ϕ are *a priori* independent. Under this assumption, the model for the response propensity can remain unspecified in order to construct valid likelihood or Bayesian inferences under the posited substantive statistical model.

Finally, missing not at random (MNAR) is reserved for all the mechanisms that are not MCAR or MAR. That is, even after blocking on the variable V, the distributions of U_{obs} and U_{mis} are different. That is, the response propensity depends upon the unknown values U_{mis} (and possibly, U_{obs} and V).

Note that both MAR and MNAR are unverifiable assumptions based on the observed data. Since U_{mis} is not known, one cannot verify that U_{obs} and U_{mis} have similar or different distributions. If one suspects the distributions to be different, then this difference has to be posited, perhaps, based on substantive knowledge or external information to conduct appropriate analysis. There are two possible approaches to model the differences, the selection model and the pattern mixture model. Analysis under these models are discussed in Chapter 7.

1.5 Problems with Complete-Case Analysis

Practitioners often ask, "Is there a level of missing data for which complete-case or available case method is reasonable?". There is no simple answer to this question as it depends upon the fraction of incomplete cases, the parameter being estimated and the amount of information available from the subjects to be discarded from the analysis. Consider a population where a variable has a "bell-shaped" distribution and the goal is to estimate the population median. In a simple random sample from this population, a fairly large amount of missing data would be needed to move the median substantially. For estimating the 90th percentile even a few missing observations can have a large impact. In a multiple regression analysis, each variable may have a small number of missing values but, collectively, a substantial number of observations may be discarded.

One of the best ways to understand the impact of missing data is through a carefully conducted simulation study. Let D be a binary dependent variable, E, a binary exposure variable and X, a single continuous covariate. Assume the following model to create these three variables:

1. $X \sim \text{Normal}(0, 1)$.

2. $E \sim \text{Bernoulli}(1, \pi(X))$ where $\text{logit}(\pi(X)) = 0.25 + 0.75 \times X$.

3. $D \sim \text{Bernoulli}(1, \theta(X, E))$ where $\text{logit}(\theta(X, E)) = -0.5 + 0.5 \times E + 0.5X$.

Suppose that the sample size is 1000. The goal is to infer about the regression coefficient in the logistic regression model (in (3) above) for D with E and X as predictors (the true value here is 0.5).

Delete some values of X with probability governed by the logistic regression model,

$$\text{logit}[Pr(X \text{ is missing})] = -1 - 0.5 \times D - 0.5 \times E + 3 \times D \times E$$

For each subject, calculate the probability based on the model, generate a uniform random number between 0 and 1 and set X to missing if this number is less than or equal to the probability. This particular model deletes values, on the average, for about 25% of the subjects but differs by D and E: 29% for $(D = 0, E = 0)$, 33% for $(D = 1, E = 0)$, 37% for $(D = 0, E = 1)$ and 6% for $(D = 1, E = 1)$. By design the data are missing at random as they depend only on the observed covariates D and E. Data are not missing completely at random as the four cell percentages of missingness are not the same. Table 1.1 provides the sample size and mean (SD) of the covariates X for the four cells formed by D and E in the before and after deletion data sets.

Clearly, the sample sizes in the four cells are different but the distribution of the covariates is similar for the before and after deletion data sets in each cell. This is expected given that the data are missing at random conditional on D and E.

A logistic regression analysis of the before-deletion data set with D as the dependent variable and E and X as predictors results in the following estimates (standard error): Intercept, $\widehat{\beta}_o$ =-0.5372 (0.1059), the coefficient for E, $\widehat{\beta}_1 = 0.4478$ (0.1420), and the coefficient for X, $\widehat{\beta}_2 = 0.5541$ (0.0782). The analysis conducted on the after-deletion data set yielded -0.5904 (0.1281), 0.8850 (0.1683) and 0.5412 (0.0922), as the respective numbers. Estimates of the intercept and the coefficient for X are similar in the before and after

Table 1.1: Descriptive statistics based on simulated before and after deletion data sets

Cell	Before Deletion		After Deletion	
	Sample Size	Mean (SD)	Sample Size	Mean (SD)
$D = 0, E = 0$	299	-0.55 (0.90)	212	-0.57 (0.90)
$D = 0, E = 1$	269	0.06 (0.85)	170	0.11 (0.82)
$D = 1, E = 0$	145	-0.13 (0.84)	97	-0.13 (0.88)
$D = 1, E = 1$	287	0.50 (0.92)	269	0.51 (0.92)

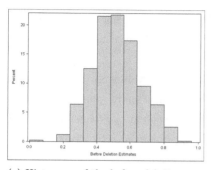
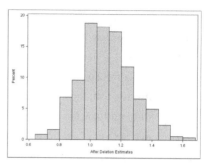

(a) Histogram of the before deletion estimates across the 500 replicates

(b) Histogram of the after deletion estimates across the 500 replicates

Figure 1.2: Results from a simulation study of logistic regression analysis with missing covariates

deletion data sets, but the estimates of the coefficient for E are remarkably different.

The simulation experiment was repeated 500 times, each time generating a complete data, fitting the logistic regression model, setting some values to missing and then fitting the same regression model for the complete cases. Figure 1.2 provides the histograms of the estimates across the 500 replications. As expected, the before-deletion estimates are centered around the true value of 0.5 whereas the complete-case estimates of severely biased and the true value is not even within the realm of 500 complete-case estimates.

If the missing data mechanism logistic model does not involve the interaction term, $D \times E$, then it has been shown that the complete-case analysis is valid (see, for example, Vach (1994)). There may be other special cases for the complete-case analysis to be valid, and it is difficult to ascertain whether those conditions are met for a particular problem. Besides, the complete-case analysis is usually less efficient. Better strategies are available for using the observed data more effectively and incorporate additional variables in the data set that may be correlated with the variables with missing values.

1.6 Analysis Approaches

The major focus of this book is on options for the analysis of incomplete data under the MAR assumption. There are two general purpose approaches for practice: weighting and imputation. Weighting approach assigns weights to subjects with no missing values to compensate for removing the subjects with missing values.

Imputation approach fills in a plausible set of values for the missing set. Multiple imputation incorporates the uncertainty in the plausible values by repeating the imputation process several times (say, M times). Each imputed set of values, when combined with the observed set of values, yields a completed data set. Each completed data set is analyzed separately. The completed data inferences are combined to form a single inference. The combining rules are relatively simple and easy to implement. Weighting or imputations involve assumptions about the nature of differences between subjects with and without missing values.

For many, imputations conjure "making-up of data." This would be true, if one were to create a single completed data set and then analyze it as though it were a complete data. The complete data cannot be created from the observed data, and the imputations are not the actual values for the nonrespondents. Collectively, however, the imputations under certain assumptions can create a plausible data set sampled from the population, thus resulting in a plausible inference about the population. This plausible inference needs to incorporate uncertainty due to imputations. The multiple imputation through multiple completed data sets incorporates this uncertainty.

The multiple imputation approach solves the missing data problem once because the same set of completed data sets can be used for a variety of analysis. This is especially appealing in a public-use setting where the data producer can use his/her knowledge, background variables and auxiliary variables to create multiple imputed (or completed data sets) and make it available to the users. The user repeats his/her analysis using his/her own complete data software on each completed data set and then combines the point estimates, standard errors, test statistics, etc. In fact, many software packages such SAS, STATA, SUDAAN and R have built-in functions to combine inferences.

Other methods can be used for incorporating the imputation uncertainty. For example, one could repeat the entire imputation process in a repeated replication setting (jackknife, bootstrap, balanced half samples, etc.) and then combine the estimates to arrive at a single inference. Under this approach, one could release all the replicated-imputed data sets to the users. The replication approach is more computer intensive but may be useful, especially if the complete data inference is going to be based on replication technique.

The third approach works directly with the statistical model and the observed data to construct inferences for the parameters. The estimation can be motivated from the likelihood principles or from a Bayesian perspective. Though this approach may be preferred from a purely theoretical perspective, its implementation may require considerable programming and analytical skills.

The goal of this book is to provide a practical guide for using weighting and imputation approaches. The emphasis is on the multiple imputation (MI) approach because it is versatile in handling many applications. Many software packages are available to implement this method. Given its wider applications, MI may not be the most efficient approach for all the problems but has numerous advantages. For example, the MI approach can be made more efficient than some of the traditional methods such as observed data maximum likelihood with effective use of auxiliary variables.

The crux of the matter, of course, is how to create imputations. The most straightforward and principled approach is to use a Bayesian framework. Start with a model for the observable, conditional on some unknown parameters, and a prior distribution for the parameters. Then, conditional on the observed data, construct the joint posterior predictive distribution of the unobserved (parameters and the set of missing values), and draw values of the unobserved from this joint distribution. The drawn values of the missing set of values are treated as imputations.

Suppose that Y is the potential data matrix with no nonresponse. Let R be the response indicator matrix of the same dimension as Y, with 1 if the corresponding element in Y is observed and 0 if it is missing. A statistical model for the observable is a joint density/mass function of (Y, R) which may involve some unknown parameters θ. Let $\pi(\theta)$ be the prior density of θ. Let Y_{obs} denote all the observed values in Y (correspond to 1s in R) and let Y_{mis} be the unknown values in Y (corresponds to 0s in R). The relevant predictive distribution for generating imputations is

$$Pr(Y_{mis}|Y_{obs}, R) = \frac{\int f(Y, R|\theta)\pi(\theta)d\theta}{\int f(Y, R|\theta)\pi(\theta)d\theta dY_{mis}} \propto \int f(Y, R|\theta)\pi(\theta)d\theta,$$

as the denominator is a function of the known values (Y_{obs}, R).

The idealistic description of the imputation process is difficult, if not impossible, to implement in practice. This book adopts the basic principle but constructs imputations through several approximations of the predictive distribution. The most prominent or emphasized approach is the sequential regression multivariate imputation (SRMI), where each variable is predicted by all others in the data set through a sequence of regression models. Many other sensible approaches are also illustrated as alternatives. To emphasize the dangers of careless imputation, nonsensible but prevalent approaches are used to illustrate the pitfalls.

The book uses actual and simulated data sets to illustrate important aspects of weighting and multiple imputation approaches. The actual data sets arise from randomized trials, observational studies and sample surveys using both longitudinal and cross-sectional study designs. This book is primarily aimed for practitioners with emphasis on discussions of the underlying assumptions, methodology and implementation. As far as the analysis of incomplete data is concerned, there are always assumptions (there is no free lunch!). It is essential to understand the assumptions for each method to judge its appropriateness or make the necessary modifications. Even though the emphasis is on applications, several sections provide theoretical basis and discussion.

It is assumed that the reader is familiar with basic statistical techniques, such as rules of probability, Bayes theorem, the basis for statistical inference, general and generalized linear models, the likelihood based inference, etc. Typically, a two course sequence in statistics or biostatistics should be sufficient to read major portions of this book. Deeper knowledge of statistical inference topics are needed to understand certain theoretical material.

Each chapter contains a section on selected readings on relevant topics. These are not exhaustive as research on missing data is vast and is nearly impossible to include the full scope of published literature. Furthermore, the original articles where the methods or concepts are developed may be highly technical and, hence, alternative references are provided. Thus, the selection of bibliography is subjective but informed by teaching students at various levels.

1.7 Basic Statistical Concepts

There are two approaches for drawing inference about a population or phenomenon: Frequentist repeated sampling and Bayesian. The philosophical difference between these approaches arise because of differing concepts of probability: its definition, construction and use.

For a practitioner, ideas from both frameworks can be useful to solve a particular scientific problem, choosing the most scientifically appropriate interpretation. This book uses ideas from both frequentist and Bayesian perspectives, but, has a definite Bayesian flavor. This section gives a very brief overview of some of the basic statistical concepts useful to understand the material in the book. A course in Bayesian analysis will be helpful.

A desire to answer a substantive research question leads to an experiment which when implemented results in data. For example, the desire to know the prevalence of diabetes in a population (the research question) leads to an experiment that involves conducting a survey of n randomly chosen subjects from the population (experiment) resulting in x people reporting having diabetes and $n - x$ people reporting not having diabetes in the sample (data). The possible values of x are $\{0, 1, 2, \ldots, n\}$ and is called the sample space (a statement of all possible results from the experiment).

A statistical model posits probability of observing a result from any arbitrary portion of the sample space. In the example, if θ were the probability that a random individual in the population has diabetes, then the probability of observing $x = 0$ is $(1 - \theta)^n$. In general, observing x diabetic subjects in the sample of size n is given by the binomial distribution,

$$f(x|\theta) = \frac{n!}{x!(n-x)!}\theta^x (1-\theta)^{n-x}$$

where the notation $n!$ stands for the product of the first n integers, $n \times (n-1) \times (n-2) \times \ldots \times 1$.

The goal of inference is to answer the following question: "Given that an experiment described above was implemented which resulted in x subjects with diabetes and $n - x$ without, what can be inferred about θ"? The two approaches, frequentist and Bayesian, differ in terms how this question is answered. Obviously θ can be any number between 0 and 1. The interval $(0,1)$ is called the parameter space (to avoid technical difficulties, assume that none being diabetic and all being diabetic are impossible scenarios for the population).

Under the frequentist framework, it is assumed that the true value of θ for this population is θ_o. A procedure is invented or developed to estimate θ_o which needs to satisfy some "good" properties across the entire spectrum of data that could result from the experiment. That procedure should then be applied to the data in hand to infer about θ_o.

The two prominent "good" properties are unbiasedness and minimum variance. Both are conceptualized based on a thought experiment. Suppose that a procedure yields an estimate $\widehat{\theta}(x)$ which depends upon x. Assume that experiment is conducted millions of times (or infinite number of times), each time resulting in x and $\widehat{\theta}(x)$. The estimate is called unbiased if the average of the estimates across the repeated experiments is equal to θ_o. This can be stated as

$$E(\widehat{\theta}(x)|\theta_o) = \sum_{x=0}^{n} \widehat{\theta}(x)f(x|\theta_o) = \theta_o.$$

The second property of minimum variance can be described as follows. Suppose that $\tilde{\theta}(x)$ is any other unbiased procedure for estimating θ_o. The variance of $\tilde{\theta}(x)$ across the repeated experiments is larger than the variance of $\widehat{\theta}(x)$. This can be stated as

$$\text{Var}(\widehat{\theta}(x)|\theta_o) = \sum_{x=0}^{n} (\widehat{\theta}(x) - \theta_o)^2 f(x|\theta_o)$$

$$< Var(\tilde{\theta}(x)|\theta_o) = \sum_{x=o}^{n} (\tilde{\theta}(x) - \theta_o)^2 f(x|\theta_o).$$

The sampling variance estimate (or simply the sampling variance) of $\widehat{\theta}(x)$ is defined as an estimate of the $Var(\widehat{\theta}(x)|\theta_o)$. The standard error is the square root of the sampling variance estimate. In the binomial example, the only "good" estimate (unbiased and minimum variance) is the sample proportion $\widehat{\theta}(x) = x/n$ and its sampling variance is $\widehat{\theta}(x)(1 - \widehat{\theta}(x))/n$.

Instead of a single estimate, it may be desirable to construct an interval that covers the true value, θ_o . A procedure results in an interval $(\widehat{\theta}_L(x), \widehat{\theta}_U(x))$ such that $Pr(\widehat{\theta}_L(x) \leq \theta_o \leq \widehat{\theta}_U(x)) \geq 1 - \alpha$ for some prespecified α. If $\alpha = 0.05$ then the interval is called a 95% confidence interval. It is preferable to be as close to equality (to $1 - \alpha$) as possible and a "good" property of the procedure is to result in a shortest possible interval for every x.

Thus, for every experiment, the idea is to construct a point and interval estimation procedure with good properties for every target quantity of interest and then apply that procedure for the particular data set in hand to construct inferences. The suffix "o" for θ may be omitted because any value in the parameter space can be the true value. Thus, the procedures should have "good" properties for any value in the parameter space.

Another type of inference is testing a preconceived hypothesis about the likely value(s) of the target quantity of interest which will not be discussed in this section but can be found in many basic textbooks (see Section 1.9). It suffices to say that, this question may also be answered by checking whether the preconceived values are included in the interval estimate.

A Bayesian accepts that there is a true value θ_o but, more importantly, the Bayesian view is that since θ_o is not known then there is *a priori* uncertainty about its value, and that should be expressed in a form of a prior distribution for θ, with the density $\pi(\theta)$ defined on the parameter space. This prior distribution could be constructed from pilot data, data from similar populations or the subject matter knowledge.

The statistical model accepted by both, frequentists and Bayesians, specifies $f(x|\theta)$ and, therefore, the product $f(x|\theta)\pi(\theta)$ is the joint distribution of (x, θ). The marginal density is $f(x) = \int_0^1 \pi(\theta, x)d\theta$, and the conditional density is,

$$\pi(\theta|x) = \frac{\pi(\theta, x)}{f(x)} = \frac{L(\theta|x)\pi(\theta)}{f(x)}, \tag{1.1}$$

where $L(\theta|x)$ is the binomial density, $f(x|\theta)$, evaluated at the observed value of x (so, it is the function of the parameter, θ) and is called the likelihood function. The conditional density given in equation (1.1) is called the posterior density of θ given the data x (and n, of course). This is the direct result of Bayes theorem in probability. This binomial problem was considered in the paper by Thomas Bayes published posthumously in 1763, through the efforts of his friend Richard Price.

The posterior density forms the basis for all Bayesian inferences about θ. This density is used to directly answer the questions of the type "What is the probability that θ is in the interval (a, b)"? The answer to this question is the integral $\int_a^b \pi(\theta|x)d\theta$.

The posterior mean defined as

$$E(\theta|x) = \int_0^1 \theta\pi(\theta|x)d\theta$$

is the average value of θ and the posterior variance

$$Var(\theta|x) = \int_0^1 (\theta - E(\theta|x))^2 \pi(\theta|x) d\theta$$

can be used to express the uncertainty.

To construct interval estimates, suppose that (a, b) is such that

$$Pr(a \le \theta \le b|x) = \int_a^b \pi(\theta|x) d\theta = 1 - \alpha$$

and for any value of θ in the interval (a, b) and θ^* outside the interval (a, b), $\pi(\theta|x) > \pi(\theta^*|x)$. That is, the values of θ inside the interval have higher values of the posterior density than for those outside the interval. The interval (a, b) is called the $100(1 - \alpha)\%$ highest posterior density credible interval. The interpretation is that the probability of the prevalence rate being between a and b is $1 - \alpha$.

For the binomial example, assume that all values of θ are equally likely, leading to a uniform prior distribution,

$$\pi(\theta) = 1,$$

for interval $(0,1)$ and is set to 0 outside the interval (as they are impossible). The posterior density can be shown to be a beta distribution with parameters $x + 1$ and $n - x + 1$,

$$\pi(\theta|x) = \theta^x (1 - \theta)^{n-x} / B(x + 1, n - x + 1)$$

with the posterior mean $\tilde{\theta}(x) = (x + 1)/(n + 2)$ and the posterior variance

$$Var(\theta|x) = \frac{\tilde{\theta}(x)(1 - \tilde{\theta}(x))}{n + 3}$$

which is not that different from the estimates from the frequentist perspective. In fact, $\pi(\theta) = \theta^{-1}(1-\theta)^{-1}$ will give results identical to the frequentist answer. However, this is not a proper density function as it is not integrable with respect to θ.

The uniform prior is equivalent to adding 1 to both the numerator and the denominator (the number of diabetic and nondiabetic subjects). A diffuse Jeffreys prior, $\pi(\theta) \propto \theta^{-1/2}(1 - \theta)^{-1/2}$, would add 1/2 instead of 1 which is shown to have better variance properties, though slightly biased.

In practice, the prior distribution is usually diffuse. Heuristically, in the binomial example, if the likelihood function and the prior density were plotted against θ, the prior density will be relatively flat when compared to the likelihood function. For a variety of problems, a diffuse prior distribution is almost equivalent to working directly with the likelihood function. It is an accepted principle that the likelihood function is the best "summary" of information in the data about the parameters, the Bayesian point of view provides the best mechanism for studying the likelihood function and constructing inferences. Thus, the Bayesian approach generally yield methods that have good frequentist properties in a broader definition of "good" procedures. Obviously, this is a point of view of a practitioner who wishes to develop easily interpretable inferences that can be justified from both philosophical perspectives.

All prior distributions in this book are diffuse in the sense that "likelihood function dominates the prior density." The diffuse priors are not always easy to define. Jeffreys prior (improper) is one example where all location parameters are treated as uniform on the real line, and the prior distribution of all scale parameters are uniform on the logarithmic scale. The prior distribution for the binomial parameter, θ, is either uniform on (0,1) or a beta prior $\pi(\theta) \propto \theta^{\alpha-1}(1-\theta)^{\beta-1}$ for small values of α and β. The main emphasis is on using the information in the data set summarized in the likelihood. Of course, methods described can be modified to incorporate information for constructing a proper prior distribution.

Modern computation methods have made simulating values from the posterior density of the parameters as a convenient and efficient way of drawing inferences. Throughout the book, this strategy is adopted as imputations are also simulation of the missing values and fits well within the Bayesian framework.

For the binomial example, the uniform prior, $\pi(\theta) = 1$, results in the posterior density of θ as a beta distribution with parameters $x + 1$ and $n - x + 1$. By simulating a large number of values from this beta distribution, many features of the posterior distribution can be studied. Furthermore, any transformation (for example, logit, $\phi = \log[\theta(1-\theta)^{-1}]$) of the simulated values are the draws from the posterior distribution of the transformed parameter. In complex problems, this feature will be useful where it may be easy to draw from one distribution but not from the other.

Extending the simple experiment, suppose that only m of n individuals answered the question about diabetes and w out of m are diabetic. In the absence of any other information, assume the data are MCAR (or MAR). Imputation involves drawing from the predictive distribution of

$$Pr(y_{m+1}, y_{m+2}, \ldots, y_n | w, m, n)$$

where $y_{m+1}, y_{m+2}, \ldots, y_n$ are the binary yes/no variables for the $n - m$ respondents. Note that,

$$Pr(y_{m+1}, y_{m+2}, \ldots, y_n | w, m, n) =$$

$$\int_0^1 Pr(y_{m+1}, y_{m+2}, \ldots, y_n | w, m, n, \theta) \pi(\theta | w, m) d\theta.$$

Now, conditional on θ, $y_j, j = m + 1, m + 2, \ldots n$ are independent Bernoulli random variables with parameter θ and θ has a beta posterior distribution with parameter $w + 1$ and $m - w + 1$. Thus, the imputation involves 2 steps:

1. Draw θ, say θ^*, from the beta distribution with parameter $w + 1$ and $m - w + 1$.

2. Draw $n - m$ independent uniform random numbers, $u_{m+1}, \ldots, u_n$ and set $y_j = 1$ if $u_j \leq \theta*$.

This two stage imputation process of drawing a value of the set of parameters and then conditional on drawn parameters, drawing the missing values is the standard strategy and is routinely used.

Now, suppose that gender (male/female) was observed on all n subjects and, under MAR, the response propensity may depend on gender. In this situation, the imputations are carried out in the same manner, except separately for males and females. When one has a large number of covariates, a regression model may be useful for creating imputations.

The frequentist concepts based on repeated sampling are useful in model checking. The thought experiment underlying the frequentist framework can be made a reality by simulating thousands of copies of data from the model. If the observed data is not within the realm of plausibility among the generated data sets (several examples will be given later in the missing data context), then the model is questionable and should be refined.

1.8   A Chuckle or Two

Many statistical properties of an inference procedure are established under the "true" model (or correctly specified model). Generally, the concept of true model is quite elusive. The thought of a proposed model being true is perhaps wrong almost all the time (a famous quote attributed to George Box is "All models are wrong but some are useful"). The concept of true response propensity is even more elusive than the prediction model. The following story, told in many classes might illustrate the point and give an opportunity to have a chuckle or two.

John, a dashing young man, just broke up with his girlfriend and was feeling depressed. He was watching TV but his mind was wandering, not concentrating on what was on the TV, as his fingers mindlessly push the buttons on the remote control. The doorbell rang and John wondered "who could that be"?

He got up and opened the front door. There was an older looking, affable gentleman who introduced himself as Peter from the University of Michigan, conducting a survey funded by a government agency. Peter explained the topic and all the usual preambles that an excellent interviewer provides to any potential respondent. John thought that this would be a great distraction and invited Peter to his living room. Peter opened his computer and administered the informed consent and thus began the interview.

John was enjoying the interview and was intrigued by the questions. Both John and Peter were developing a good rapport and the survey interview was progressing nicely. Suddenly the phone rang, startling both John and Peter. John, saw that it was from his girlfriend (ex?), Jody. He said, "Excuse me, let me take this phone call, I will be right back." He left the living room to be in private and while walking he heard Jody sobbing. She was mumbling "I am so sorry, so sorry. . . ." John couldn't make out what she was saying and said, "Honey, why are you crying? What is the matter"? His heart was thumping loudly. Jody composed herself, and said "I am sorry; I shouldn't have broken up with you. I want to give another chance to our relationship. Can you come here right now, please"? John melted. Now John felt Peter's company as a nuisance!! John said, "Sure honey, I will be there in few minutes."

John rushed to the living room frantically, he said to Peter, "Something very urgent has come up. I have to leave right now. Let us reschedule the interview for a later date." Peter pleaded that it will only take about 10 more minutes but to no avail. John and Peter settled to meet about a week later. Peter was literally pushed out the door with a polite thank you for understanding.

John locked up the apartment and rushed to Jody's place which was about a 20-minute drive. They both were so thrilled to see each other, and they talked endlessly for hours. Finally, Jody said "We need to do something to strengthen our relationship." John replied, "I agree. What do should we do"? Jody after pondering for a few seconds said "You should just move in today"! John didn't even blink an eye and said "Yes."

They went to John's apartment, packed up boxes and suitcases with essential items and came back to Jody's place. John had forgotten all about Peter, the interview and the rescheduled date. Both Jody and John were immersed in building their relationship and giving a strong footing. Poor Peter. He showed up at John's apartment on the scheduled appointment date. Peter tried to locate John by calling his landline several times without success. He tried to find information about John from neighbors, and they all said they hadn't seen John in a while and didn't know his whereabouts.

Peter reluctantly marked John as a partial responder. Peter was about to ask John some income and wealth questions when the ominous phone call from Jody came.

Now what response propensity model will capture why John had missing values for the income and wealth questions? There may be many stories behind the reasons for nonresponse. No logistic regression model (or any other regression model) can capture all nuances. The best that a practitioner can do is to match John based on what is known about him with Bill who responded and looks like John in many respects and make a best effort possible to predict for John (or weight Bill by two to compensate for missing John).

For a practitioner, it is important to understand the structures in the data and develop models that appears reasonable. If he/she is not sure about the model, it is important to perform sensitivity analysis to the stated model assumptions. Use simulation to check the properties of inferences to deviation in the model assumption.

1.9   Bibliographic Note

As one can imagine, the problem of missing data is probably as old as the data analysis itself. Elderton and Pearson (1910) used mean imputation for missing covariates that was criticized by Sturge and Horsely (1911). Pearson (1911) provides painstaking care to address the limitation of the data in hand and also justification of their methods. See Stigler (1999) for various other debates about the paper by Elderton and Pearson (1910) and Pearson (1910, 1911) which describe perhaps the first concrete application of statistics to social sciences.

Kermack and McKendrick (1927, 1932, 1933) developed a series of equations relating the observed and target (unobserved or missing) variables as a part of theory of epidemics. McKendrick (1926) is perhaps the first systematic presentation of recursive estimation across a variety of problems which may be called a version of the EM-algorithm, systematically developed in Dempster, Laird and Rubin (1977).

One of the finest developments in statistics, the design of experiments, is a carefully structured data collection approach to evaluate the effect of various factors on an outcome. The structure placed emphasis on making the effects orthogonal or independent of each other which also made the analysis simple. The beauty of such structure is destroyed when one or more values of the outcome is missing. Various methods were proposed to estimate the missing values, usually, in an iterative fashion and adjusting the degrees of freedom for the reduction in the sample size. The clever methodology was first described in Allan and Wishart (1930) and later generalized by Yates (1933). Bose and Mahalanobis (1938) and Bose (1938) consider the problem estimation when only the linear combination of variables are available (mixed-up yields from an agricultural experiment). Dodge (1985) provides further details about the evolution of methods for analyzing designed experiments with missing values. Most of the methods were driven by computational considerations, less of an issue in modern times.

Wilks (1932) develops a method of moments estimates and their sampling distribution for the means, variances and the correlation coefficient based on a sample with missing data from a bivariate normal distribution. Anderson (1957) developed the maximum likelihood estimates of parameters in the multivariate normal distribution when the sample had some missing values. The paper exploits the monotone pattern of missing data to develop noniterative solutions to the score equations. Hartley (1958) considers a more general case of maximum likelihood estimation from incomplete data. Hartley and Hocking (1971) and Orchard and Woodbury (1972) provide prelude to fully developed EM-algorithm in Dempster, Laird and Rubin (1977). This approach and the increasing computing power has led (and continues) to numerous refinements to handle complex models. Implementing them in practice usually involves writing computer codes tailored to a specific model/situation.

Glasser (1964) was perhaps the first to explicitly introduce the response indicators in the context of a regression analysis with missing covariates, and defined missing at random as the independence of missing data indicators across the variables. Afifi and Elashoff (1966, 1967) provide a summary and literature review of missing observations on multivariate statistics.

Though Glasser (1964) introduced the response indicators, an explicit model by treating them as random variables, its dependence on the study variables was proposed by Rubin (1976) which actually transformed the landscape of models and methods for analyzing incomplete data. The model for response indicators led to a more principled approach for evaluating the assumptions and development of appropriate methods. Another seminal contribution by Rubin (1978a) is multiple imputation, a general framework for analyzing data with some missing values.

Weighting approach has its origins in the sample survey literature to compensate for unit nonresponse. It is still heavily used in the sample surveys. Hartley (1946), Politz and Simmons (1949), Deming (1950) and Simmons (1954) are some of the early references that use weighting to adjust for nonresponse. This approach can be used for some item nonresponse situations and is discussed in Chapter 2.

As indicated earlier, numerous ad hoc approaches for handling missing data have been proposed. Jones (1996) evaluates the use of indicator variable for missing variables in a linear regression model and Vach (1994) for a logistic regression model. Both show that this strategy can result in biased estimates and should be avoided. Little (1992) provides a more comprehensive review of methods for regression analysis with missing covariates.

A classic book by Little and Rubin (1987, 2002) is the first to provide a systematic treatment of analysis of incomplete data. Rubin (1987) is a classic for the multiple imputation approach. Chapter 2 in Rubin (1987) provides the foundation for the model based analysis of survey and nonsurvey data with missing values. Schafer (1997) provides another excellent read on missing data. Recent books, Carpenter and Kenward (2013) and van Buuren (2012), explore multiple imputation methods in more details. Molenberghs et al (2014) is a compendium of missing data methodology to date.

The basic textbooks for a review of the frequentist approach are Hogg McKean and Craig (2012), Casella and Berger (2002) and Cox and Hinkley (1979). For an overview of the Bayesian analysis, one of the excellent books is by Gelman et al (2013). Box and Tiao (1973) is a classic and very useful for people to learn basics of Bayesian analysis.

1.10 Exercises

1. Consider a regression problem with Y as a continuous dependent variable with two predictors, X_1 and X_2. The predictor X_1 and the dependent variable Y are fully observed. Some values of X_2 are missing. Let R_2 be the response indicator taking the value 1 for respondents and 0 for nonrespondents. State whether the following mechanisms are MCAR, MAR or NMAR.

   (a) $\text{logit}[Pr(R_2 = 1|X_1, X_2, Y)] = \beta_o + \beta_1 X_1 + \beta_2 Y$

   (b) $Pr(R_2 = 1) = 0.2$

   (c) $R_2 = 1$ if $Y \geq y_o$ for some known y_o

   (d) $\text{logit}[Pr(R_2 = 1|X_1, X_2, Y)] = \beta_o + \beta_1 X1 + \beta_2 X_2$

   (e) $\text{logit}[Pr(R_2 = 1|X_1, X_2, Y)] = \beta_o + \beta_1 X_1$

   (f) $R_2 = 1$ if $X_1 \geq x_{1o}$ for some known x_{1o}

2. A survey is being conducted based on a random sample of firms from the population list which has the name and size of the firm (number of employees). A key survey variable of interest is whether or not the firm offers health insurance to its employees and the number of health plans offered. Consider the following missing data mechanisms.

   (a) All firms exceeding some certain firm size refuse to participate.

(b) Firms that do not offer health plans and/or very limited number of plans are more likely to be nonrespondents.

State whether the plans are MCAR, MAR or NMAR. What are the challenges in performing analysis under each mechanism?

3. Many studies collect data on time-to-event also called survival time, failure times, etc. Subjects may drop out of the study so the time-to-event is not fully observed and such observations are called censored observations. Relate this problem to the missing data mechanism and explain what MCAR, MAR and NMAR means in terms of differences between censored and noncensored observations.

4. **Project**. Repeat the simulation study described in Section 1.5 and also under the following missing data mechanism:

$$\text{logit}[Pr(X \text{ is missing})] = -1.5 + 0.25 \times D + 0.45 \times E$$

Compare the scatter plots for the two mechanisms, and analyze the impact of including the interaction term $D \times E$ in the missing data mechanism.

5. **Project**. One way of understanding the effect of missing data mechanism on the complete case analysis is to conduct a simulation study. Generate one hundred data sets under the following model assumptions:

(a) $X_2 \sim N(0,1)$;

(b) $X_1 \sim N(X_2, 1)$; and

(c) $Y \sim N((X_1 + X_2)/2, 1)$.

Perform the following analysis on each data set and save the five parameter estimates (before deletion estimates):

(a) Mean of X_2

(b) Regression of Y on X_1 and X_2

(c) Regression of X_2 on Y and X_1

Now delete some values of X_2 using each mechanism (1) to (5) in Problem 1. Choose values of the regression coefficients, x_{1o} and y_o to yield about 40% missing values. On each data set with missing values perform the same analysis and save the five parameters (complete-case estimates).

For each mechanism and each parameter, construct a scatter plot of 100 pairs of estimates (before deletion, complete-case). Note which scatter plots look close to the 45 degree line. Based on this analysis write a brief report on the impact of missing data mechanisms on each analysis.

2

Weighting Methods

2.1 Motivation

The idea behind weighting is simple: to make the respondents as similar as possible to the original sample in terms of the distribution of some variables. Suppose that in a sample of size 100, there were 60 women and 40 men. Among the 60 women, suppose 40 provided information on income, while only 10 among the 40 men provided income. Clearly, the respondents providing income have lost their original representation on gender. If the analysis is restricted to respondents, then bias may be introduced, especially if the income is correlated with gender. To compensate or correct for the loss of representation, a weight of $60/40 = 3/2 = 1.5$ is attached to each responding woman and a weight of $40/10 = 4$ is attached to each responding man. With these weights, the respondents are now weighted back to the original sample representation of 60 women and 40 men. The numbers 1.5 and 4 are called nonresponse adjustment weights and the two cells formed based on gender are the adjustment cells.

In a probabilistic framework, 40 women respondents can be viewed as a subsample of 60 women with the selection probability 2/3, and 10 men as a subsample 40 men with the selection probability 1/4. That is, 50 respondents have been subsampled from 100 subjects in the original sample with varying probabilities of selection, and the weights are proportional to the inverse of these probabilities. These probabilities are called response probabilities or propensities. Women have a response propensity of 2/3 and men, 1/4.

A study design yielding the observed data may be viewed as an approximate two-phase sample design, where in the first phase 100 subjects were selected and asked gender. In the second phase, a subsample of women and men were selected with the stated probabilities and income was ascertained. The two-phase or multiphase design is a very common in many large scale surveys. These designs have appeared in econometric literature and in biostatistics in more recent times. The two-phase approximation is useful for creating appropriate inferences.

There is one major difference between the regular two-phase design and the approximation discussed above. The selection probabilities are fixed *a priori* in the two-phase survey designs across all possible samples from the population whereas in the nonresponse situation, the second phase selection probabilities could vary from sample to sample. That is, the selection probabilities are random variables as they are subject to sampling variation. The standard error calculations based on weighted data should reflect this additional uncertainty or sampling variability in the weights. However, many software packages analyzing weighted data ignore this additional variability by treating the weights as fixed numbers.

An implicit assumption underlying the weighting procedure in the example is that within the gender group, the distributions of income for respondents and nonrespondents are the same. That is, the data are missing at random, conditional on the gender. Specifically,

$$Pr(\text{income is missing}|\text{gender, income}) = f(\text{gender}).$$

Alternatively, the equality of the distribution of the income variable between respondents and nonrespondents within the same gender category can be expressed as:

$$Pr(\text{income}|\text{gender} = g, \text{respondent}) = Pr(\text{income}|\text{gender} = g, \text{nonrespondent})$$

On the other hand, the missing completely at random assumes that the distributions of income for the entire 50 respondents and the entire 50 nonrespondents are the same (ignoring the gender). Though neither assumption can be verified, MAR assumption is less stringent (involving separation by gender) than the missing completely at random assumption.

The goal of weighting, therefore, is to create homogeneous groups based all available variables measured on both respondents and nonrespondents and then weight the respondents in each group to represent the full sample in that group. Several variables may be available on respondents and nonrespondents and one needs to create homogeneous groups based on these variables.

2.2 Adjustment Cell Method

Consider the data presented in Table 1.1. There are four adjustment cells defined by D and E. The number of respondents is 212 and nonrespondents is 87 in the cell $D = 0, E = 0$. The weights for each of the 212 respondents will be $w_1 = (212 + 87)/212 = 1.4104$. The weight for 170 respondents in the $D = 0, E = 1$ is $w_2 = (170 + 99)/170 = 1.5198$. Similarly the weights for the 97 respondents in the cell $D = 1, E = 0$ is $w_3 = 1.4949$ and for $D = 1, E = 1$ is $w_4 = 1.0669$.

A weighted logistic regression (for example, PROC SURVEYLOGISTIC in SAS) can be fit to the complete cases to compensate for differential response rates across these 4 cells. The weighted estimates and their standard errors are $\widehat{\beta}_o =$-0.5306 (0.1281), $\widehat{\beta}_1 = 0.4287$ (0.1693) and $\widehat{\beta}_2 = 0.5477$ (0.0903). The weighted estimates are close to the before-deletion data set estimates. The standard errors are obviously larger than the complete-data standard errors.

The adjustment cells, generally, are formed by creating a cross-classification of variables measured on both respondents and nonrespondents. Continuous variables are categorized. With many variables, the cross classification may create a sparse table and, hence, highly variable weights. The cells have to be collapsed to make the estimated response rate stable enough to construct the weights. A random effect model could be used to estimate the cell-specific response rates by smoothing or shrinking the cell-specific estimates towards the overall estimate. Thus, the development of nonresponse adjustment weights is problem specific. A general purpose approach for developing weights is based on the propensity score methods.

2.3 Response Propensity Model

Consider an observational study to evaluate the effect of a binary treatment or exposure variable on the outcome. In the absence of randomization, the propensity scores are often used to create balanced groups of treated and untreated subjects, and then the distribution of the outcome variable is compared across these homogeneous groups.

The situation is similar here where the response indicator R is the "treatment" or "exposure" (with $R = 1$ denoting respondents and $R = 0$ for nonrespondents) and X is set of covariates available for respondents and nonrespondents. A response propensity is the probability of $R = 1$ expressed as a function of X.

Considerable literature in survey methodology is devoted to understand the nature of the relationship between R and X. If one were interested in trying to increase the response rates, variables X that can be subjected to experimentation (intervened) is of considerable interest. Thus, the study of relationship between R and X is of considerable interest on its own merit for developing design modifications to improve the response rates.

The goal of adjustment, however, is different. Attempt is being made to reduce nonresponse bias in the inference "after-the-fact." This is similar to the analysis of observational study where R is a treatment or exposure indicator and has not been randomized except that in the missing data problem the "null hypothesis" of no difference between the respondents (treated or exposed) and nonrespondents (untreated or unexposed) is assumed to be true (MAR conditional on the response propensity). It is important therefore to include in X, not only the variables that explain differences between $R = 1$ and $R = 0$, but also make the null hypothesis of no differences in $f(Y|R = 0, X)$ and $f(Y|R = 1, X)$ plausible. Thus, the focus of the selection of X should be on those that are related to both R and Y. In other words, the goal is to achieve independence between R and Y conditional on X.

Once the covariates $X = (X_1, X_2, \ldots, X_p)$ have been identified, a regression model will be fit with R as the dependent variable and X as independent variables. This regression model can be fully parametric such as logistic or probit regression models, semiparametric such as the generalized additive model or nonparametric such as the classification and regression tree (CART).

To be concrete, consider a logistic regression model,

$$\text{logit}[Pr(R = 1|X)] = \beta_o + \beta_1 X_1 + \beta_2 X_2 + \ldots + \beta_p X_p,$$

where $\beta_o, \beta_1, \ldots, \beta_p$ are the unknown regression coefficients. These unknown parameters are obtained using the maximum likelihood approach which leads to an estimated propensity score,

$$\hat{p} = [1 + \exp(-\hat{L})]^{-1}$$

where $\widehat{L} = \widehat{\beta}_o + \widehat{\beta}_1 X_1 + \ldots + \widehat{\beta}_p X_p$ is the linear predictor. The propensity score from a well fitting regression model is the coarsest scalar summary of the covariates that can be used to assess the imbalance between the respondents and nonrespondents.

Finding the best fitting propensity score (the logistic regression) model is an iterative process beginning with a candidate model, checking the balance on each covariate given the estimated propensity score, refining the model, rechecking, etc. The importance of these step cannot be understated. Here are some techniques that can be used to check the balance of the covariates, conditional on the estimated propensity scores:

1. Suppose that a covariate, say, X_1 is a continuous variable. Regress X_1 on $\widehat{p}$ (or $\widehat{L}$) and obtain the residual e_1. Compare the distributions of e_1 for the respondents and nonrespondents. If they overlap and look similar then the propensity score balances the distribution of this covariate between the two groups (respondents and nonrespondents). Histograms and kernel densities of residuals are useful to assess the similarity and overlap. One can also perform formal tests for the equality of distributions of the residuals for the two groups. Often, these tests may show statistical significance, especially when the sample size is large and has poor power when the sample size is small. If the distributions of the residuals are not similar, then the model needs to be refined. It could be due to some nonlinearity (for example, a scatter plot of e_1 versus X_1, may be useful in this context) or may need an additional interaction term between X_1 and another variable.

2. For a binary covariate, create four cells $C_1 = (X_1 = 0, R_1 = 0)$ or $C_2 = (X_1 = 0, R_1 = 1)$, $C_3 = (X_1 = 1, R_1 = 0)$ and $C_4 = (X_1 = 1, R_1 = 1)$. Check for the similarity of the distributions of the propensity scores for the cells C_1 versus C_2 and also C_3 versus C_4. A box-plot may be used to graphically inspect the equality of the propensity scores. Alternatively, run a logistic regression of X_1 on $\widehat{p}$, and obtain deviance residuals e_1. Check for the similarity of the distribution of the residuals between respondents and nonrespondents.

3. For a count variable, one can use a Poisson regression with X_1 as the dependent variable and $\hat{p}$ as the independent variable. Again, the deviance residuals can be used to check for the balance between respondents and nonrespondents.

Though the forgoing discussion was based on using the logistic regression framework, there are other alternatives, such as probit, classification and regression trees (CART), and generalized linear regression models with other links (log, negative binomial distribution, etc.). The goal is to construct a summary of the covariates that achieves the balance of covariates between respondents and nonrespondents.

2.4 Example

As an example, consider inference about the population distribution of self-reported health from a survey. For the weighting procedure to be effective, covariates are needed on both respondents and nonrespondents, and they need to be correlated to both R and Y. In practice, however, obtaining such covariates needs planning at the design stage. Interviewer observations or measures collected during the data collection process (**paradata**) might be one source of such covariates. The second source might be administrative data providing either individual level data or neighborhood level data and, finally, variables in the sampling frame or a list from which sample may have been drawn.

Assets and health dynamics of old Americans (AHEAD) is a national probability face-to-face survey of subjects who are at least 70 years of age. Interviewers were asked to collect several observations while trying to recruit subjects for an interview. The four binary variables were whether or not subjects made statements that fall into the following four categories:

1. Time delay statements such as professing lack of time to do surveys or needing to attend to urgent matters, etc.

2. Negative statements about participation in the survey such as some previous bad experiences with interviews, surveys are not that useful, etc.

Table 2.1: Mean (SD) or proportion for the six individual level variables by response status

Variable	Respondents	Nonrespondents	Effect Size
Age	77.41 (6.81)	77.83 (6.12)	0.065
Percent female	63.3	65.6	0.006
Percent making time delay statements	10.2	16.1	0.175
Percent making negative statements	9.5	37.7	0.703
Percent making positive statements	11.6	2.0	0.388
Percent mentioning old age or illness	8.1	13.8	0.183

3. Positive statements about participation in the survey, civic responsibilities, etc.

4. Statements about being old or unable to participate for one reason or another

The following were the neighborhood level variables assembled from a variety of administrative sources or constructed based on the interviewer observations:

1. Large urban area (yes/no)

2. Barriers to contact (yes/no)

3. Block level: Persons per sq-mile

4. Block level: % persons 70 or older

5. Block level: % Minority populations

6. Block level:% Multi-unit structures (10+)

7. Block level: % Occupied housing units

8. Block level: % Single person housing units

9. Block level: % Vacant housing units

10. Block level: Persons per occupied housing unit

In addition to these fourteen variables, age and gender were also available on all sampled subjects. The original sample size was $n = 10,173$ and the number of respondents was $m = 8,212$ giving the response rate of 80.7%.

The last column in the above table is the standardized difference between the respondents and nonrespondents defined as

$$D = \frac{|\bar{x}_R - \bar{x}_{NR}|}{\sqrt{(s_R^2 + s_{NR}^2)/2}}$$

Table 2.2: Mean (SD) of ten housing or block level variables by response status

Variable	Respondents	Nonrespondents	Effect Size
% From large urban areas	15.4	22.7	0.19
% With barriers to contact	12.7	20.0	0.20
Persons per square mile	6162.2 (14318.6)	7750.9 (15086.2)	0.11
% Persons 70 or older	10.98 (8.18)	11.16 (7.96)	0.02
% Minority populations	22.76 (30.16)	24.60 (31.94)	0.06
% Multiunit structures (10+)	10.96 (20.24)	12.65 (21.74)	0.08
% Occupied housing units	33.50 (23.76)	35.43 (24.68)	0.08
% Single person housing units	24.48 (11.54)	25.17 (11.75)	0.06
% Vacant housing units	9.05 (9.18)	8.82 (8.90)	0.03
Persons per occupied housing unit	2.64 (0.47)	2.62 (0.48)	0.03

where $\bar{x}_R$ is the mean for the respondents, $\bar{x}_{NR}$ is the mean for the nonrespondents, s_R is the standard deviation for the respondents and s_{NR} is the standard deviation for the nonrespondents. For a binary variable, $\bar{x}$ is the proportion, and the standard deviation is $s = \sqrt{\bar{x}(1-\bar{x})}$. D is a commonly used measure in the propensity score analysis of the observational studies with a binary treatment variable. For practical purposes, $D \leq 0.25$ may be termed small, $0.25 < D \leq 0.5$ medium, $0.5 < D \leq 0.75$ large and $D > 0.75$ as very large effect sizes. Thus, in the above table, there are 4 small, 1 medium and 1 large effect sizes.

Table 2.2 provides the mean (SD) or proportion for the neighborhood level variables by response status and the corresponding effect sizes. Effect size for all the neighborhood level variables are small but do indicate that there are some meaningful differences in the neighborhood characteristics of the respondents and nonrespondents. For instance, to construct approximate z test to assess the statistical significance of the differences between the respondents and nonrespondents, the effect sizes may be divided by $\sqrt{1/8212 + 1/1961} = 0.025$. Thus, any effect size larger than 0.05 will be deemed statistically significant at the 5% level of significance. However, the problem with using statistical significance is that with a large sample size, even the small effect sizes may be declared as significant. The descriptive statistics in Tables 2.1 and 2.2 clearly show that the data are not missing completely at random as many of the variables are related to the response propensity.

To construct nonresponse adjustment weights, fit a logistic regression model with the response indicator for the self-reported health variable as the dependent variable and the sixteen variables as predictors. The goal is to obtain a well fitting model that balances the covariates between respondents and non-respondents. One useful measure of goodness of fit is the Hosmer-Lemeshow type test where deciles based on the estimated propensity score from the current model are created and then the observed and expected frequencies /counts in these 10 classes are compared using a chisquare statistics with 9 degrees of freedom as a measure of lack of it.

Once satisfied with the logistic model, create classes (or strata or groups) based on quartiles or quintiles or even deciles (depending upon the sample size) of the estimated propensity score. Developing the propensity score model and checking its balancing property is an iterative process.

One can start with the main effect model and check the Hosmer-Lemeshow chisquare statistic. At the second stage, add two factor interactions to reduce the value of chisquare statistic. One can also use stepwise or another variable selection procedure to reduce the model complexity. Higher order interactions may be added, if necessary. As in any regression analysis, the covariates may be transformed or the continuous covariates may be standardized (i.e., subtract the overall mean and divide by the overall standard deviation) for obtaining a better fitting model.

For the AHEAD example, the first model with all 16 main effects resulted in the Hosmer-Lemeshow chisquare statistic of 28.68 with the p-value of 0.0004. Next, the model with all two factor interactions was fitted which reduced the chisquare statistic as 17.33 with the p-value of 0.03. As is typical, the full interaction model may be overmatching or fitting too many parameters resulting in a poor fit.

The next step was to use a stepwise selection with different entry and exit probabilities. After several tries, the procedure with 0.5 as the cut-off for both entry and exit for the variables in the stepwise selection, obtained the model with chisquare statistic of 4.75 with the p-value 0.784. In any regression analysis, there are multiple strategies for developing a prediction model and the procedure described above is not the only approach. The ultimate goal is to reduce the discrepancy between the observed and fitted frequencies across the entire range of the estimated propensity scores. For example, instead of using stepwise, one could use ridge regression or lasso to obtain the prediction model and the propensity scores.

Four strata were created based on quartiles of the estimated propensity scores from the final model and the effect sizes for all sixteen variables were

Table 2.3: Nonresponse adjustment weights based propensity score stratification

Propensity Score Strata	Respondents	Nonrespondents	Weights
1	1505	1037	1.6890
2	2100	445	1.2119
3	2237	306	1.1368
4	2370	173	1.0730

computed in each strata. The largest effect size was 0.03 indicating a good balance between respondents and nonrespondents conditional on the propensity score from this model. One simple way to construct weights is to use the four strata as adjustment cells and attach weights to the respondents in each cell. Table 2.3 provides nonresponse adjustment weights.

An alternative approach is the inverse probability weighting, where the weight for a respondent is the inverse of his/her response propensity, $w_i = 1/\widehat{p}_i$, $i = 1, 2, \ldots, m$. This can be viewed as an extreme stratification where each respondent constitutes a stratum. Figure 2.1 shows the histogram of the nonresponse adjustment weights.

The survey variable of interest is the self-reported health on a five point scale: excellent, very good, good, fair, and poor. Table 2.4 provides weighted

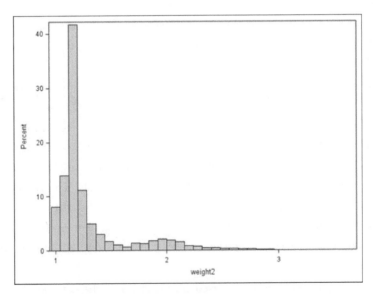

Figure 2.1: Histogram of the non response adjustment weights

Table 2.4: Unweighted and weighted frequency distributions (in %) and their standard errors of self-reported health

Self-Rated Health	Unweighted	Propensity Score Strata Weights	Inverse Probability Weighting
Poor	10.75	10.51 (0.34)	10.51 (0.34)
Fair	22.8	22.53 (0.47)	22.61 (0.47)
Good	30.35	30.41 (0.52)	30.50 (0.52)
Very good	23.08	23.29 (0.48)	22.31 (0.48)
Excellent	13.03	13.26 (0.38)	13.06 (0.38)

and unweighted frequency distributions and their standard errors. Once the weights have been constructed, the data needs to be analyzed using the survey analysis software package. The standard error in Table 2.4 was obtained using the SAS software package, PROC SURVEFREQ. The survey analysis packages are available in other software environments such as STATA or R as well.

Practically, there are no differences between unweighted and the two versions of the weighted frequency distributions. The variables used in the weight construction though correlated with the response indicator but are not correlated with the outcome variable. This can be seen from the following box plot (Figure 2.2) comparing the distribution of weights across the five categories of the outcome variable.

It is important to select variables to be included in the response propensity models. The main criterion is that they be related to both R and Y. Generally, the weighted estimates have larger standard error, especially if the response propensity scores are poorly correlated with the outcome.

2.5 Impact of Weights on Population Mean Estimates

Suppose that the weights are normalized so that they add to the number of respondents m. The difference between the weighted and unweighted mean of the outcome y is

$$b = \sum_i^m (w_i - 1)y_i/m = \sum_i u_i y_i/m$$

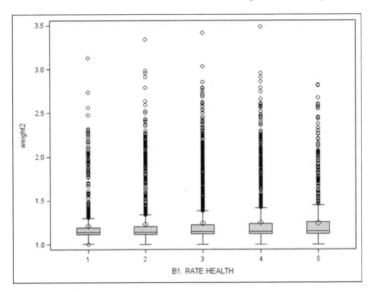

Figure 2.2: Box plot of the response propensity weights by self-rated health status

where $u_i = w_i - 1$. In expectation, the difference will be approximately the sum of the covariance between U and Y (or the covariance between W and Y) and the product of the mean of U and Y. The mean of U is approximately zero (as the weights are normalized to the sample size). That is, for large samples, b may be considered as the sample covariance between u and y. Thus,

$$E(b) \approx Cov(U, Y) = \rho_{WY}\sigma_W\sigma_Y$$

where ρ_{WY} is the correlation between W and Y, and σ_W and σ_Y are the standard deviations of W and Y, respectively. The standardized difference $E(b)/\sigma_Y$ is approximately $\rho_{WY}\sigma_W$. An estimate of this quantity is $r_{wy} \times \sqrt{\sum_i w_i^2/m - 1} = r_{wy}c_w$ where r_{wy} is the sample correlation coefficient between w and y, and c_w is the coefficient of variation of the weights. For example, if the correlation coefficient is 0.3 and the coefficient of variation of the weights is 0.1, then the standardized difference will be 0.03, which is a small effect size. If the coefficient of variation is 0.5 and the correlation coefficient is 0.8, then the effect size is 0.4.

The coefficient of variation of the weights also plays a role in increasing the sampling variances. The sampling variance of b can be approximated as

$$Var(b) = \frac{m-1}{m^3}\left(E(U^2)E(Y^2) + (m-1)E(U^2Y^2) - (m-2)E^2(UY)\right)$$

which reduces to

$$\frac{m-1}{m^2}(1 + \rho_{WY}^2)\sigma_W^2\sigma_Y^2$$

when U and Y are approximately normally distributed. Thus, large variance of the weights result in large uncertainty in b or b/σ_Y.

Further assuming that $1 - m^{-1} \approx 1$, the effect of weighting can be assessed using the z-score,

$$z = \frac{\sqrt{m}(b - r_{wy}s_w s_y)}{\sqrt{(1 + r_{wy}^2)s_w^2 s_y^2}}$$

where s_w and s_y are the sample standard deviations for the weights (normalized to the number of respondents) and the response variable, respectively. Two critical quantities that govern the effect of weighting are the correlation coefficient between the weight and the response variable and the coefficient of variation of the weights (or the standard deviation of the normalized weights). Since the weights are not that highly correlated with the response variable, it is clear that the weighted estimates will have a larger variance for a relatively small reduction in bias.

2.6 Post-Stratification

2.6.1 Post-Stratification Weights

Nonresponse adjustment methods discussed so far use the covariate information from respondents and nonrespondents to create weights for the respondents to make them representative of the full sample. Sometimes aggregate information may be available about the population from which the sample has been drawn. The numbers on demographics from the most recent census is an example of such external information. To be concrete, consider the following table with age and gender composition of the respondents in the behavior risk factor surveillance system for Michigan in 2010 (top number in the cell) and the population size from the 2010 census (the second number in the cell).

Table 2.5: Construction of post-stratification weights

Variables	Age Category					
	18-24	25-34	35-44	45-54	55-64	65+
Male	170	181	371	652	871	1,130
	494,025	580,833	633,321	743,803	608,637	587,184
	2,906.03	3,209.02	1,707.05	1,140.80	698.78	520.16
Female	144	318	610	1,012	1,342	2,062
	479,864	583,316	644,653	766,230	643,360	774,346
	3,332.39	1,834.33	1,056.81	757.14	479.40	475.53

The goal of post-stratification is to assign weights to respondents in each cell to make them representative of the population size for that cell. For example, 170 males between the ages of 18 to 24 in cell (1,1) will be assigned a weight $w_{11} = 494,025/170 = 2906.03$. The third number in each cell gives the post-stratification weights to be assigned to each respondent in that cell. The underlying assumption is that the data are missing at random conditional on the cells used in forming the post-stratification cells.

2.6.2 Raking

Sometimes, the cell-specific population size is not be available but the marginal totals are published. A procedure called raking is used to develop weights that satisfy the marginal totals. It is an iterative procedure of row and column margin allocations until convergence is achieved. Suppose that, for the data given in the previous table, only the marginal population sizes of males and females or for the age categories are available:

Table 2.6: Post-stratification example with raked weights

Variables Age Category	Males		Females		Sample Total	Population Total
	Count	Weights	Count	Weights		
18-24	170	3591.94	144	2522.63	314	973,889
25-34	181	2879.19	318	2022.06	499	1,164,149
35-44	371	1598.66	610	1122.74	981	1,277,974
45-54	652	1108.10	1,012	778.22	1,664	1,510,033
55-64	871	693.14	1,342	486.79	2,213	1,256,997
65+	1,130	528.10	2,062	370.89	3,192	1, 361,530
Sample Total	3,375		5,488		8,863	
Population Total	3,647,803		3,896,769			7,544,572

The raking in this situation assumes that the post-stratification weight, w_{ag} for the cell with age $= a, a = 1, 2, \ldots, 6$ and gender $= g, g = 1, 2$ is of the form $\alpha_a \beta_g$ such that $\sum_a w_{ag} n_{ag} = N_{+g}$ the marginal population size for gender g and $\sum_g w_{ag} n_{ag} = N_{a+}$ the marginal population size for age category a. An implicit missing data mechanism under this approach is $Pr(R = 1|\text{age} = a, \text{gender} = g) = Pr(R = 1|\text{age} = a)Pr(R = 1|\text{gender} = g)$.

The eight quantities, $\alpha_a, a = 1, 2, \ldots, 6$ and $\beta_g, g = 1, 2$ are determined using an iterative process starting with an adjustment for one margin and then to the second, revisiting the first and then the second and so on until the convergence is achieved. Specifically, the constraint $\sum_a w_{ag} n_{ag} = N_{+g}$ results in the equation $\beta_g \sum_a \alpha_a n_{ag} = N_{+g}$ or $\beta_g = N_{+g}/\sum_a \alpha_a n_{ag}$. The constraint $\sum_g w_{ag} n_{ag} = N_{a+}$ results in the equation $\alpha_a = N_{+a}/\sum_g \beta_g n_{ag}$. Suppose that $\alpha_a^{(t)}$ and $\beta_g^{(t)}$ values at iteration t, the following iterative equations can be used to estimate the post-stratification weights:

$$\alpha_a^{(t)} = \frac{N_{a+}}{\sum_g \beta_g^{(t-1)} n_{ag}},$$

and

$$\beta_g^{(t)} = \frac{N_{+g}}{\sum_a \alpha_a^{(t)} n_{ag}}.$$

The raked post-stratification weights are given in Table 2.6 when only marginal population sizes are available on age and gender categories. Starting with any arbitrary values of β or α, the iterative equations converge in about three steps. These weights slightly differ from those given in Table 2.5 which are based on the population joint distribution of age and gender.

The raking concept can be extended to many variables and many kinds of situations. Suppose that the population sizes for the 12 age-gender categories are available and the marginal population sizes for four categories of race/ethnicity (non-hispanic white, non-hispanic black, hispanic and other) are available. In this case, the post-stratification weights are assumed to be of the form $w_{agr} = \gamma_{ag} \delta_r$ (or implicitly assume that

$$Pr(R = 1|\text{age} = a, \text{gender} = g, \text{race} = r) =$$

$$Pr(R = 1|\text{age} = a, \text{gender} = g)Pr(R = 1|\text{race} = r).$$

The 16 unknown quantities in the weight definition can be estimated using an iterative process similar to the one described above that matches on the 12

Table 2.7: Sample proportions reported not having any health insurance

Post-Stratification	Age Category					
Variables	18-24	25-34	35-44	45-54	55-64	65+
Male	29	27	19	19	12	2
Female	23	20	17	13	13	2

two-way margin totals $N_{ag+}, a = 1, 2, \ldots, 6, g = 1, 2$ and 4 race margin totals, $N_{++r}, r = 1, 2, 3, 4$.

More recently, the idea of post-stratification has been extended to develop what are called calibration weights. Suppose that population totals $T_1, T_2, \ldots, T_p$ for p variables $X_1, X_2, \ldots, X_p$ are known. Suppose that these variables are also measured in a survey with n respondents. The calibration weights $w_i, i = 1, 2, \ldots, n$ attached to the respondents should satisfy the constraints, $\sum_i w_i x_{ji} = T_j$ for $j = 1, 2, \ldots, p$ where x_{ji} is the value of the variable X_j measured on subject i. Obviously, there are more unknowns than equations or constraints as n will be much larger than p. Some additional criterion such as minimizing the variance of the weights subject to those p constraints may be used to develop the calibration weights. The strategy could be used to develop a weighted estimate of the population total for an outcome variable Y using the calibration weights that minimize the mean-square error. The former criterion is a general purpose weight to be attached an individual, and the latter is more tuned towards inference for a particular variable Y.

2.6.3 Post-stratified Estimator

The following Table 2.7 provides percentage of respondents who reported not having any kind of health coverage including health insurance, prepaid plans such as HMO, or government plans such as Medicare or Indian Health Service in Michigan 2010.

Using the post-stratification weights from Table 2.7 results in the weighted estimate as

$$\frac{\sum_a \sum_g w_{ag} p_{ag}}{\sum_a \sum_g w_{ag}} = \frac{29 \times 2906.03 + \ldots + 2 \times 475.03}{2906.03 + \ldots + 475.03} = 21.12,$$

and using the raked post-stratification weights from Table 2.7 results in the weighted estimate as

$$\frac{29 \times 3591.94 + \ldots + 2 \times 370.89}{3591.94 + \ldots + 370.89} = 21.32.$$

These two can be compared to the unweighted estimate of 10.91. Obviously, the unweighted estimates are severely biased, given the large differences between the sample and population distributions of age and gender, and patterning of insurance status by age and gender.

The sampling variance of the post-stratified estimator is computed by combining sampling variances of stratum specific estimates. Noting that $var(p_{ag}) = p_{ag}(1 - p_{ag})/n_{ag}$, the sampling variance of the post-stratified estimator is

$$\frac{\sum_a \sum_g w_{ag}^2 p_{ag}(1 - p_{ag})/n_{ag}}{(\sum_a \sum_g w_{ag})^2}. \qquad (2.1)$$

The above formula assumes that the weights are constant, but actually they are not because the cell sizes may vary from sample to sample. Some have suggested that this extra uncertainty should be included in the variance calculations and others have argued that this extra variation is unrelated to the information in the data for constructing the estimates of the parameter and hence, one should assume these weights as fixed constants. Nevertheless, the extra variation is often negligible for large sample sizes.

Suppose that n is the sample size, $W_{ag} = w_{ag}/\sum_a \sum_g w_{ag}$ are the normalized weights (that is, add up to 1), the extra variation due to nonconstant cell size is

$$\sum_a \sum_g W_{ag}(1 - W_{ag})p_{ag}(1 - p_{ag})/n_{ag}/n.$$

For the estimated proportion of people who reported not having any insurance, the sampling variance is 1.18×10^{-5} or the standard error $\sqrt{0.000118} = 0.0109$ (or 1.09% in terms of the percentage points) and added variance due to random cell sample sizes is 7.06×10^{-8}, a small quantity relative to the sampling variance. Using the raked weights, the estimated standard error is 1.08%. Knowing the population cell count or just the population margins lead to similar inferences but different from the unweighted analysis.

The example used in this section is for illustrating the basic concepts of post-stratification and relevant calculations. The actual BRFSS data provides more refined weights that include sampling, nonresponse adjustment and post-stratification weights. The final weight is the product of all three weights.

The methods described in this chapter can be used to create additional weights, for example, when cases are deleted due to missing values in the covariates when fitting a regression model. Consider fitting a regression model with health status (H) score as the dependent variable, age (A), gender (G), and health insurance coverage (I) as independent variables. Suppose that W is the weight variable in the BRFSS database and missing values are only in I. A response propensity weight to compensate for deleting subjects with missing values can be derived as follows. Let $R = 1$ if I is observed and 0 if I is missing. Fit a regression model to estimate $Pr(R = 1|A, G, H, W)$ (for example, the logistic or probit regression model). The final weight for the respondents with no missing values in A, G, I and H will be $W/\widehat{Pr}(R = 1|A, G, H, W)$.

2.7   Survey Weights

In many sample surveys, the original design may involve unequal probabilities of sampling which results in sampling weights W_S. The unit nonresponse adjustment weights (W_{NR}) and post-stratification weights (W_{PS}) may also be constructed to account for nonresponse. The final weight to be used in the analysis is the product,

$$W_F = W_S \times W_{NR} \times W_{PS}$$

which is designed to make the respondent pool representative of the target population of interest. Often the survey agencies use variables from a variety of sources to construct these weights and many such variable are not included as a part of the released data set. Thus, these weights could be functions of variables that are correlated with several survey variables. Ignoring the weights can result in bias as well as loss of efficiency.

The AHEAD example is illustrative of weights not having much information correlated with the particular survey outcome (possibly an increase in variance) and the BRFSS example is illustrative of weights highly relevant for reducing the bias. Generally, the concern is about bias but pay the price of an increase in sampling variance. Thus, it is always prudent to do a weighted analysis.

During the course of weight construction, one can encounter some extreme weights. This can happen if the covariate distribution is sparse and logistic regression, while extrapolating, may yield a very low probability of response. Since the weights are the reciprocals of the probability of response, weights can become very large. In such cases, individuals with large weights may dominate the analysis. The weights need to be trimmed, which may introduce bias. A careful analysis of weights, as if it is one of the study variables, is necessary to avoid problems with the results. Creating classes based on weights rather than treating them as a continuous variable avoids such problems. Additional references on weight trimming are provided in Section 2.11.

2.8   Alternative to Weighted Analysis

Suppose that the original design was a simple random sample of size n from a large population. As before, suppose that $w_i, i = 1, 2, \ldots, m, m + 1, \ldots, n$ are the weights for the m respondents and $n - m$ nonrespondents. Without loss of any generality. the first m observations correspond to respondents and the last $n - m$ for the nonrespondents. Suppose that $y_i, i = 1, 2, \ldots, m$ are the response variables on the m respondents. The prediction approach fills the data on the nonrespondents and then constructs relevant estimates. The standard error then incorporates variances due to sampling of respondents and also the prediction error for the nonrespondents.

Suppose that the empirical analysis shows that the following regression model fits the data well,

$$y_i = \beta_o + \beta_1 g(w_i) + h(w_i)\epsilon_i$$

where $\epsilon_i \sim iid\ N(0, \sigma^2)$ and g and h are known functions. The least squares prediction for the nonrespondents is $\widehat{y}_i = \widehat{\beta}_0 + \widehat{\beta}_1 g(w_i)$, $i = m+1, m+2, \ldots, n$. The sample mean with the filled-in data,

$$\widehat{\mu} = \frac{\sum_i^m y_i + \sum_i \widehat{y}_i}{n},$$

is an estimate of the population mean, μ. Under the correctly specified model, $\widehat{y}_i$ is unbiased for $E(y_i|w_i)$ and, since w_i is fully observed, the predictions recover the mean of the joint distribution of (y, w) and, thus, the mean of the marginal distribution of y.

The standard error calculation is slightly complicated. Suppose that $D = \{(y_i, w_i), i = 1, 2, \ldots, m; w_i, i = m + 1, \ldots, n\}$ is the observed data. We have,

$$Var(\widehat{\mu}) = Var[E(\widehat{\mu}|D)] + E[Var(\widehat{\mu}|D)].$$

The first term is the complete data variance as computed from the filled-in data, and the second term is the additional variation due to prediction. For simple regression models, these two terms can be calculated analytically. An alternative is based on a bootstrap approach as described below.

1. Draw a sample of size n with replacement from the original sample of bivariate pairs $\{(y_i, w_i), i = 1, 2, \ldots, m; (., w_i), i = m + 1, m + 2, \ldots, n\}$ where "." indicates the missing value.

2. Fit the regression model on this sample and predict the missing values.

3. Calculate the statistic, t, of interest from the filled-in sample.

4. Repeat steps 1, 2 and 3 several, say B, times.

The bootstrap estimate is $\bar{t}_B = \sum_i^B t_i/B$ and its standard error is given by

$$SE^2(\bar{t}_B) = \sum_i^B (t_i - \bar{t}_B)^2/[B(B-1)].$$

Incidentally, the weighted mean, $\sum_i w_i y_i / \sum_i w_i$, can be viewed as the least squares estimate of the regression coefficient, β, from the complete-case analysis based on the following regression model, $y \sim N(\beta, \sigma^2 w^{-1})$. The advantage of the prediction approach is the flexibility in developing the model and incorporation of weights for nonrespondents in the analysis. However, the weights for the nonrespondents are not usually available or released as part of the public use data, making this approach difficult to implement.

The following modification of the bootstrap approach could be implemented when the weights for the nonrespondents are not available:

1. Draw a sample of size $n - m$ subjects with replacement from n subjects and assign their w_i to create $\{(., w_i), i = m+1, m+2, \ldots, n\}$.

2. Draw a sample of size n with replacement from the augmented sample of bivariate pairs $\{(y_i, w_i), i = 1, 2, \ldots, m; (., w_i), i = m + 1, m + 2, \ldots, n\}$ where "." indicates the missing value.

3. Fit the regression model on this sample and predict the missing values.

4. Calculate the statistic, t, of interest from the filled-in sample.

5. Repeat steps 1, 2, 3 and 4 several, say B, times.

6. Calculate the bootstrap estimate and its standard error as before.

2.9 Inverse Probability Weighting

Over the last several years, inverse probability weighting methods have been extended that use both prediction using a regression model and a correction term using inverse probability weighting. For subject i, suppose that y_i is the response variable subject to missing values, and x_i are the predictors with no missing values. The goal is to estimate $E(y) = \mu$ and assume that the data are missing at random. The regression model $E(y|x) = h(x, \beta)$ is used to construct the prediction $\widehat{y}_i = h(x_i, \widehat{\beta})$ where $\widehat{\beta}$ is the ordinary least squares estimate obtained by minimizing the sum of squares, $\sum_i^m (y_i - h(x_i, \beta))^2$, where m is the number respondents, and n is the sample size. Let w_i be the weight (reciprocal of the predicted response propensity) based on a response propensity model $\pi(x, \alpha) = Pr(R = 1|X, \alpha)$. A class of estimator of the form,

$$\widehat{\mu} = \sum_i^n \widehat{y}_i/n + [\sum_i^m w_i(y_i - \widehat{y}_i)]/n,$$

has been proposed. This class of estimator is termed doubly robust because they are consistent estimators when at least one of the models (prediction or the response propensity) is correctly specified. The effort is needed to make sure that the analyst makes a reasonable attempt to construct well fitting models for both prediction and response propensity to minimize the impact of model mispecification.

The major emphasis in this book is on prediction modeling but use the the response propensity models merely as a device to obtain a balancing score for the weight construction or grouping respondents and nonrespondents into homogenous groups. In fact, the later chapters emphasize prediction modeling for the imputation purposes and keep the response propensity model unspecified or arbitrary, as it is very difficult to conceive a "true" response propensity model.

2.10 Bibliographic Note

The idea of weighting originated probably in various government agencies in the 1930s. Some of the early work was mentioned in Chapter 1. Brick (2013) provides a review of weighting approach for unit nonresponse. Many books have appeared that discuss weights in a broader context of the analysis of data from complex surveys. Some examples are Graubard and Korn (1999), Chambers and Skinner (2003), Herringa, West and Berglund (2010), Lumley (2010) and Valliant, Dever and Kreuter (2013).

Post-stratification is a general purpose strategy to improve representativeness of the sample (even with 100% response rate) but it is used in the context of unit nonresponse. Holt and Smith (1979) and Little (1993) are excellent readings on post-stratification. Kalton and Flores-Cervantes (2003) is also an excellent overview of the weighting methods.

Readings for calibration weights should include Deville et al (1993), and Sarndal (2007). Much of calibration weights methodology can be found in Sarndal, Swensson and Wretman (2013). This is an evolving area and numerous papers are appearing in the literature. Some of these methods can be implemented in software packages such as R, SAS and Stata.

Raking involves fitting log-linear models by setting nonestimable parameters to 0 and constraining to the known margins. The iterative proportional fitting algorithm described in Bishop, Fienberg and Holland (2007) may be used to develop the post-stratification weights. This has been implemented in many packages. A publicly available SAS macro CALMAR2.SAS (Sautory (2003)) and RAKINGE.SAS (Izrael, Hoaglin and Battaglia (2000)) can be used to develop weights in the SAS software environment. Stata and R routines are also available to create calibration and post-stratification weights.

Some early works on trimming weights include Potter (1988, 1990). For model based approaches for weight trimming, see Elliott and Little (2000) and Elliott (2007, 2008, 2009).

The basic idea of prediction approach for handling weights described in this book is a simplification of ideas in Zheng and Little (2003). The inverse probability weighting methods in a broader context will be discussed in Chapter 5. Several refinements of this basic idea of prediction using survey weights is an evolving area of research.

The original reference for inverse probability weighting method is Robins et al (1994). Many refinements have taken place (see Cao, Tsiatis, and Davidian (2009)) and Seaman and White (2013) provide a review of inverse probability weighting for missing data.

The bootstrap approach for estimating the standard errors is based on Efron (1994).

For more details about AHEAD, see Soldo et al (1997).

2.11 Exercises

1. **Project.** Download the 2009-10 data from the National Health and Nutritional Examination Survey (NHANES) available from the National Center for Health Statistics (www.cdc.gov/nchs/nhanes.htm). Extract the following covariates: age, gender, race, ethnicity and education. Also extract hemoglobin A1c from the laboratory portion of the data files. Ignore the missing values in the covariates as well as the complex survey feature of NHANES. Recode, race-ethnicity variable into a four category variable: 1. non-hispanic whites, 2. non-hispanic blacks, 3. hispanics and 4. other. Recode education into a four category variable: 1. less than high school 2. high school 3. some college and 4. completed college. The goal is to estimate the population mean of hemoglobin A1c level.

   (a) Compute the complete-case estimate of the mean.

   (b) Develop adjustment cell weights and the weighted estimate of the mean.

   (c) Develop Response Propensity weights and the weighted estimate of the mean.

   (d) Analyze the correlation between the response propensity or adjustment cell weights with the outcome variable and assess the magnitude of the effect of weighting on the estimates.

2. **Project.** Continuing from the previous project, download tables from the Census Bureau web site on population counts based on age, gender, race/ethnicity and education. Obtain tables as cross-classified as possible. Develop a strategy for calculating the post-stratification weights. Construct the post-stratified estimator of the population mean and its standard error. Discuss the effect of post-stratification weight in relation to the effect of nonresponse adjustment weights.

3

Imputation

An imputation based approach involves replacing the set of missing values with a plausible set of values. This plausible set is created or estimated based on available information. This task should not be construed as creating "actual" or "real" values for nonrespondents. For a layman, this idea conjures an image of a statistician making up the data. This is true only if one were to analyze the data as if the imputed data are real values. Thus, any single imputation analysis that does not account for the imputation process in the ensuing statistical inferences may inflate the precision and declare effects significant when they are not, and thus invalid.

Consider a simple example to assess the severity of this problem. Suppose that, in a sample of size 100, 70 respondents provided the data yielding a respondent sample mean of 50 and a sample standard deviation of 10. Under the missing completely at random assumption, the estimate of the population mean is 50 and the standard error is $10/\sqrt{70}$. Suppose that the observations on 30 subjects are filled in by drawing values at random with replacement from the 70 subjects. The filled in data, when plugged into any software package, to estimate the mean and its standard error will yield the mean of 100 observations (70 original and 30 filled-in) to be 50, on average and the standard error as the standard deviation of the filled in data (which on average will be 10) divided by $\sqrt{100}$. Of course, this is incorrect because the 30 observations filled are not adding any information that is not already there in the 70 observations. In fact, in the absence of any other information, the correct standard error from the filled-in data set has to be somewhat larger than $10/\sqrt{70}$ as we are adding noise to the observed data. Thus, the analysis of singly imputed data, without reflecting the uncertainty about the imputed values is incorrect and should be avoided in practice.

This chapter discusses approaches for creating plausible values for the missing set of values and Chapter 4 discusses construction of correct inferences from the data set with imputed values. The major emphasis is on multiple imputation. An alternative approach on replicating the imputation process using bootstrap or jackknife techniques is discussed in Chapter 9.

The goal of imputation is to fill in the missing set of values with **plausible values** to obtain a **plausible sample data** from the population. This can then be used to construct **plausible inferences** about the population quantities of interest. The notion of plausibility can be best explained heuristically. Consider a scenario with two variables Y_1 and Y_2 in a data set with Y_1 fully observed and Y_2 with some missing values. Assume that the data are missing at random. That is, the missing data mechanism depends only upon Y_1. Suppose that the missing values in Y_2 have been imputed using some procedure. How does one check heuristically whether the imputed values are plausible from the population?

A simple approach is to construct a scatter plot based on the **completed data set** (or the filled-in data set) with Y_1 on the horizontal axis, Y_2 on the vertical axis and with observed and imputed values differentiated using colors or symbols. For a plausible data set from the population (under the stated assumptions), the observed and imputed values should be interchangeable. The idea of plausible data set rules out imputing the mean of the observed values (unconditional mean) or the predicted values from a regression equation. These methods may be useful for specific purposes, but they are not general purpose imputation methods.

Suppose that in a sample of size n measuring a variable Y, only m values are observed. The mean of the observed values, $\bar{y}_o$, is imputed for every $n - m$ missing values. Assume that the data are MCAR (or MAR in the absence of any covariates). In this case, the complete-case analysis is correct. The mean of the completed data is $\bar{y}_o$ which is the correct mean. The complete data variance will be inflated as the numerator is still $\sum_i^m (y_i - \bar{y}_o)^2$, but the denominator is $n - 1$. That is, the completed-data variance is smaller than s_o^2, the variance based on m observed values; the completed data variance is $s_C^2 = s_o^2(m - 1)/(n - 1)$. Depending upon the lack of symmetry among the observed values, the median, mode and other percentiles also will be incorrect (that is, different from the complete-case analysis).

Now, consider a problem with two variables Y_1 and Y_2 with Y_1 fully observed, and Y_2 is observed on m out of n individuals. Assume that the data are missing at random. A linear regression model may be developed and the predicted values from this model may be considered as imputation. The observed values would have error around the regression line, whereas the imputed values do not. Again, the variance computed from the completed data will be smaller affecting other statistics. Statistics that are not affected are the estimated regression coefficients (intercept and slope).

Sometimes the mean imputation is used (rather inappropriately) in the scale construction. Suppose that a scale is defined as summing a set (say, k) of yes/no (or Likert scale items) items. Not all scale items are observed. An arbitrary rule is made: For a given subject, if more than k_o items are missing then his/her scale value is set to missing; otherwise, substitute the mean of the observed item responses for the missing items and then proceed with the scale construction. At the face value, there are two things wrong with this approach. First, it ignores some observed values on a set (may even be a large number) of subjects. Second, it makes the assumption that there will be no variation in the responses of the items that are missing. This artificially decreases the variation in the scale and affects all analyses that involves the spread of the scale (i.e., correlation and regression analysis).

Substitution of mean or predicted values was developed in the era of limited computational power and for situations where the tabulation of means and proportions is of interest such as reports from the government agencies and other official statistics documents. These methods should be avoided in practice now given the methodological developments, vastly improved computational power and software environment. In the later chapters, pitfalls of using such methods will be illustrated through examples and homework problems.

3.1 Generation of Plausible Values

The most straightforward approach to generate a plausible completed-data set is through draws from the predictive distribution of the missing set of values, conditional on the observed set of values (and external information, if any). For the simple bivariate problem, a set of plausible values may be generated as follows:

1. Fit a regression model $Y_2 = f(Y_1) + \epsilon$ and define $\widehat{Y_2} = \widehat{f}(Y_1)$ for both respondents and nonrespondents.

2. Construct residuals for the respondents, $\hat{\epsilon} = Y_2 - \widehat{Y_2}$.

3. For each nonrespondent, draw a random residual from the respondents and add it to his/her predicted value.

The above approach, though works in practice, does not account for all the uncertainties involved in the construction of the predictive distribution. A principled way to generate plausible values is to use a Bayesian framework. The following modification of the above procedure incorporates the uncertainties:

1. Suppose that $I = \{1, 2, \ldots, m\}$ are the indices of the respondents (that is, both Y_1 and Y_2 observed). Draw a sample of size m from the index set I with replacement and denote it as I^*.

2. Sample m values from the index set I^* with replacement and extract the corresponding values of Y_1 and Y_2.

3. Carry out the steps (1), (2) and (3) in the simple approach discussed previously.

Steps (1) and (2) in the modified approach is the approximate Bayesian bootstrap (ABB) and incorporates the uncertainty in estimating the prediction equation and in the distribution of the residuals. This procedure uses a mix of parametric and nonparametric assumptions to construct the predictive distribution. It uses a parametric model for the functional relationship but arbitrary or unspecified distribution for the residuals.

Now consider the following procedure. Create C strata based on Y_1. Suppose that n_c and m_c are the sample size and the number of respondents, respectively, in cell $c = 1, 2, \ldots, C$. Randomly sample $n_c - m_c$ observations from the m_c observations as imputation set for that cell. This has nonparametric flavor where empirical distribution in each cell is used as the predictive distribution. This is the basic setup of the hot-deck method.

A fully parametric model may be used, if appropriate. Suppose that the following regression model, $Y_2 = \beta_o + \beta_1 Y_1 + \epsilon$ with $\epsilon \sim N(0, \sigma^2)$ fits well to the respondent data. Assume that the prior information is diffuse or noninformative as indicated before, $Pr(\beta_o, \beta_1, \log \sigma) \propto c$. Let $\hat{\beta} = (\hat{\beta}_o, \hat{\beta}_1)^t$ be the 2×1 vector of regression coefficients, $V = X^t X$ is the 2×2 matrix where X is $m \times 2$ matrix with the first column of 1s and the second column Y_1 and

$\widehat{\sigma}^2$ is the estimated residual variance. The following steps result in plausible values for the missing set:

1. Draw a chi-square random variable, u, with $m-2$ degrees of freedom and define $\sigma_*^2 = (m-2)\widehat{\sigma}^2/u$. This step draws a value from the posterior distribution of σ^2.

2. Let T be the Cholesky decomposition of V such that $TT^t = V$. Let $Z = (z_1, z_2)^t$ be a 2×1 vector of independent standard normal deviates. Define $\beta^* = \widehat{\beta} + \sigma_* TZ$. This step draws from the posterior distribution of β conditional on the observed data and σ_*.

3. For a nonrespondent, define the imputed value as $Y_2^* = \beta_o^* + \beta_1^* Y_1 + \sigma_* z_3$, where z_3 is another independent standard normal deviate. Repeat the process for all nonrespondents (each time drawing a new z_3). This step draws from the posterior predictive distribution of Y_2 given the observed data, β^* and σ^*.

This approach assumes only the observed data $[(Y_2, Y_1)$ on the respondents and Y_1 on the nonrespondents] as known and all the missing values in Y_2 and model parameters β_0, β_1 and σ^2 as unknown, drawing values from their joint posterior distribution, conditional on the observed data. This approach is implemented in many missing data software packages.

Thus, there are many ways of constructing the predictive distribution. Some are described as algorithms and others are based on explicit model assumptions. Underlying all the algorithms, there is an implicit model. This model needs to be gleaned from the algorithm and checked against the data. That is, the model needs to be "validated" through proper model diagnostics for obtaining valid inferences from the imputed data. Careless imputations can yield biased estimates (even more biased than the complete-case estimates).

3.2 Hot Deck Imputation

The predictive distribution can be created based on explicit or implicit models as explained in the previous section. Hot deck is a popular approach for creating imputations using an implicit model. To be concrete, consider the data in Table 1.1 with four cells based on D and E with missing values in

X. Just like in the weighting procedure, consider the four cells as imputation cells.

A hot deck procedure creates imputations by sampling with replacement the needed number of observations from the respondent pool within each cell. For example, in the cell with $D = 0, E = 0$, sample 87 observations with replacement from 212 respondents. An implicit model is that, for each cell, the predictive distribution for the missing observations is the "empirical distribution" estimated based on the respondents. The empirical distribution function can be expressed as

$$\widehat{F}_{de}(x) = \sum_{i}^{m_{de}} I_{[x_i \leq x]}/m_{de},$$

where m_{de} is the number of respondents in the cell $D = d, E = e; d, e = 0, 1$, and $I_{[x_i \leq x]} = 1$ if $x_i \leq x$ and 0 otherwise is the indicator function.

If the number of respondents in each cell is small, then some smooth estimate of $\widehat{F}$ can be constructed or some cells can be collapsed. This procedure closely aligns with the adjustment cell weighting method. Here instead of assigning weights to the respondents, the values are imputed for the nonrespondents by drawing values from the respondents.

There are many other ways of constructing hot deck imputation cells. For example, stratification based on the propensity score can form imputation cells. In the AHEAD example, the quartiles based on the propensity scores may be considered as imputation cells, the values for the nonrespondents in any given cell can be sampled from the respondents from the same cell.

Another popular approach for creating imputation cells is based on the predicted value of the variable with missing values (predictive mean matching). Suppose that U is the variable with some missing values, and V is the set of predictors with no missing values. Regress U on V using an appropriate, good fitting model based on the respondents. To construct the prediction equation one should use proper exploratory and diagnostic techniques to arrive at a reasonable model. Suppose $\widehat{U}_i$ is the predicted value for subject $i = 1, 2, \ldots, n$, where n is the full sample size. Note that U is observed for the respondents, but $\widehat{U}$ can be computed for the full sample. Create cells based on $\widehat{U}$, and then sample an appropriate number of observations with replacement from the respondent pool in each cell. Unlike the methods based on the cross-classification of V or the propensity score stratification, this approach uses the observed data on U as well.

There are no fixed rules in terms of how to construct the cells. The goal is to make the respondents (called donors of their responses) and nonrespondents (called recipients of the values from the donors) as similar as possible on the variables measured on both. One could use the response propensity, predicted values and any other combination of variables to create imputation cells. This flexibility in using the observed data makes the approach quite useful in practice.

3.2.1 Connection with Weighting

Suppose that C imputation cells or strata have been formed with n_c and m_c as the sample size and respondent size, respectively, in stratum $c = 1, 2, \ldots, C$. Let $y_{ic}, i = 1, 2, \ldots, m_c; c = 1, 2, \ldots, C$ be the observed responses. Let $\bar{y}_c$ be the sample mean in cell c. The weights for observations in cell c is $w_c = n_c/m_c$ and the weighted estimate is $\bar{y}_w = \sum_c w_c \bar{y}_c/n$. The weighted mean can be rewritten as

$$\bar{y}_w = \sum_c \sum_i^{m_c} w_{ic} y_{ic}/n$$

where $w_{ic} = n_c/m_c$.

Instead of weighting, suppose that the missing values are imputed by drawing $n_c - m_c$ values from the m_c respondents in each cell. Let r_{ic} be the number of times y_{ic} was selected in the imputation process. The imputation or completed-data estimate is

$$\bar{y}_I = \sum_c \sum_i (1 + r_{ic}) y_{ic}/n.$$

In cell c, $n_c - m_c$ draws are made and each of the m_c observations has a chance of $1/m_c$ of being drawn at each draw. Since the draws are independent, $E(r_{ic}|D) = (n_c - m_c)/m_c$ or $E[(1+r_{ic})|D] = w_{ic}$ where D is the observed data, $\{n_c, m_c, y_{ic}, i = 1, 2, \ldots, m_c; c = 1, 2, \ldots, C\}$. The expected value of $\bar{y}_I$ over repeated imputations of the missing values, conditional on the observed data is $\bar{y}_w$. Thus, across repeated imputations and conditional on the observed data, the hot deck estimate will be distributed around the weighted mean (that is, $E(\bar{y}_I|D) = \bar{y}_w$).

Conditional on the observed data, the variance of $\bar{y}_I$ is

$$Var(\bar{y}_I|D) = \frac{1}{n^2} \sum_c \sum_i y_{ic}^2 Var[(1 + r_{ic})|D].$$

Noting that $Var[(1 + r_{ic})|D] = (n_c - m_c)m_c^{-1}(1 - m_c^{-1})$. Some algebraic simplification obtains

$$v_D = Var(\bar{y}_I|D) = \sum_c \left[\frac{n_c - m_c}{n^2} \{(1 - m_c^{-1})^2 s_{yc}^2 + (1 - m_c^{-1})\bar{y}_c^2\} \right),$$

where $\bar{y}_c$ and s_{yc} are mean and the standard deviation of the observations in cell c.

Thus, the unconditional variance of $\bar{y}_I$ is

$$Var(\bar{y}_I) = Var[E(\bar{y}_I)D)] + E[Var(\bar{y}_I|D)]$$

$$= Var(\bar{y}_w) + E(v_D).$$

It is analytically difficult to compute $E(v_D)$ since it is a nonlinear function of the random variables $n_c, m_c, \bar{y}_c$ and s_{yc}^2. It is of the order n^{-1} as can be seen using the following approximation. Assume that $1 - m_c^{-1} \approx 1$, hence,

$$v_D = \frac{1}{n} \left(1 - \frac{m}{n}\right) \sum_c p_c(s_{yc}^2 + \bar{y}_c^2)$$

where $p_c = (n_c - m_c)/(n - m)$, $n = \sum_c n_c$ is the sample size and $m = \sum_c m_c$ is the number of respondents. In this situation, there is no benefit in using the hot deck imputation estimates over the standard weighted estimate as it is less efficient and randomly varies from the weighted estimate.

3.2.2   Bayesian Modification

From a Bayesian point of view, all uncertainties need to be incorporated to obtain valid inferences. In the hot deck procedure, the imputations are drawn from the estimated (or empirical) distribution without incorporating uncertainty in that estimate. A simple modification uses the Bayesian bootstrap in each imputation cell as follows:

1. Draw $m_c - 1$ uniform random numbers between 0 and 1 and order them to obtain $u_o = 0 \le u_1 \le u_2 \dots \le u_{m_c-1} \le 1 = u_{m_c}$.

2. Draw a uniform random number, v, and choose y_{ic} as the imputation if $u_{i-1} < v \le u_i$.

3. Repeat Step 2 to fill all $n_c - m_c$ missing values in the cell.

4. Repeat the procedure for all the cells.

The underlying theory is that the distinct values in the set $\{y_{ic}, i = 1, 2, \ldots, m_c\}$ are modeled as having a multinomial distribution with unknown cell probabilities. These probabilities are assumed to have a noninformative Dirichlet distribution. The above procedure is equivalent to obtaining draws from the posterior predictive distribution of the missing values conditional on the observed values under this model assumption.

An alternative approach is to draw values using the approximate Bayesian bootstrap procedure as follows:

1. Sample m_c values with replacement from the m_c values of $y_{ic}, i = 1, 2, \ldots, m_c$.

2. Sample $n_c - m_c$ with replacement from the sample obtained in step 1.

3. Repeat the process for all the cells.

The process for creating imputation cells is similar to developing adjustment cells for weighting. All model building and diagnostic procedures need to be applied to create sufficiently large and homogenous groups so that missing at random (conditional on the imputation cells) becomes plausible. The variables used to form imputation cells needs to be correlated with the variables being imputed.

The hot deck procedure described so far focusses on imputing a single variable, conditional on fully observed covariates on respondents and nonrespondents. In practice, however, several variables may have missing values and the number of fully observed covariates may be limited or even nonexistent. Thus, a sequential or a multivariate approach may have to be used to impute all the missing values as described later.

3.3 Model Based Imputation

Let y_i be the complete data vector on subject $i = 1, 2, \ldots, n$ which is assumed to be sampled from the population that follows a distribution with the density function $f(y_i|\theta, z_i)$, where z_i is a vector of auxiliary variables with no missing values. The statistical model is explicitly stated (i.e., parametric such as multivariate normal, log-linear model, etc. or semiparametric such as generalized additive models etc. or nonparameteic such as classification and regression trees, etc.). Let r_i denote a vector (of the same dimension as y) of response indicator variables with 1 if the corresponding element in y_i is observed and 0, if missing. Let $y_{i,obs}$ denote the components of y_i that are observed (corresponding elements in r_i are all equal to 1) and $y_{i,mis}$ denote components of y_i that are missing (corresponding elements in r_i are all equal to 0). The notation can become a bit confusing because r_i is used as a random variable defining the missing data mechanism and also as an indexing set to identify which variables are observed or missing.

The missing data mechanism is the probability specification $g(r_i|y_i, z_i, \phi)$, governing which components in y_i are observed or missing. Thus, the full statistical model is the joint distribution of (y_i, r_i), $f(y_i, r_i|z_i, \theta, \phi)$, which can be decomposed into $f(y_i|z_i, \theta)g(r_i|y_i, z_i, \phi)$. This is called a selection model decomposition. On the other hand, the joint distribution can be decomposed into $h(y_i|r_i, z_i, \alpha)k(r_i|z_i, \beta)$, which is called a pattern-mixture model decomposition. In practice, sometimes the selection model decomposition is easier to formulate and may be useful from a conceptual point of view, and other times the pattern-mixture model may be more appealing. Obviously, when both are defining the same joint distribution, there is a mapping,

$$f(y_i|z_i, \theta)g(r_i|y_i, z_i, \phi) = h(y_i|r_i, z_i, \alpha)k(r_i|z_i, \beta.).$$

Sometimes the mapping may be useful to understand the proposed missing data mechanism, under a pattern mixture model, or to understand the pattern mixture model under the proposed selection model decomposition. The difference between the two formulations will play a crucial role under the NMAR mechanism, discussed in Chapter 7.

Under MAR mechanism, $g(r_i|y_i, z_i, \phi) = g(r_i|y_{i,obs}, z_i, \phi)$, and ignorability condition (that is, ϕ and θ are not functionally (from a frequentist perspective) or probabilistically (from a Bayesian perspective) related to each other), the predictive distribution is

$$f(y_{i,mis}|y_{i,obs}, z_i, r_i, \theta, \phi) \propto f(y_{i,mis}|y_{i,obs}, z_i, \theta).$$

Thus, only the substantive model is needed for imputation purposes and the missing data mechanism can be unspecified (as long as ignorability condition is assumed to be true).

Obviously, θ is not known. If $\widehat{\theta}$ is an estimate then the imputations can be drawn from $f(y_{i,mis}|y_{i,obs}, z_i, \widehat{\theta})$. This approach ignores the uncertainty in the estimate $\widehat{\theta}$. A proper procedure is defined below:

1. Let $\pi(\theta)$ be the prior density for θ.

2. Let $y_{obs} = \{y_{i,obs}, i = 1, 2, \ldots, n\}$, $z = \{z_i, i = 1, 2, \ldots, n\}$. Construct the posterior density

$$\pi(\theta|y_{obs}, z) \propto \pi(\theta) \times \prod_i^n f(y_{i,obs}|z_i, \theta)$$

where $f(y_{i,obs}|z_i, \theta) = \int f(y_i|z_i, \theta) dy_{i,mis}$ is the marginal distribution of the observed portion of the complete data.

3. Draw a value θ^* from $\pi(\theta|y_{obs}, z)$ and then draw imputations from $f(y_{i,mis}|y_{i,obs}, z_i, \theta^*)$ and thus incorporating uncertainty about θ.

The idealized description may not be achievable in practice. There are many ways to approximate this strategy (some were discussed for the simple bivariate example). Some general methods are listed below:

1. Let $\widehat{\theta}$ be the mode of the posterior distribution and $\widehat{V}$ be the second derivative of the log posterior density. This is easy to implement using the Newton-Raphson method or other numerical optimization techniques, available in SAS, STATA and R packages for various statistical models. Approximate the posterior density of θ by a multivariate normal density with mean $\widehat{\theta}$ and $(-\widehat{V})^{-1}$ as the variance. Draw θ^* from this normal distribution and then draw from $f(y_{i,mis}|y_{i,obs}, z_i, \theta^*)$ to create imputations.

2. When the sample size is small or in the case of skewed situations (such as logistic regression with rare events), the multivariate normal distribution may not be a good approximation. Importance ratios may be used to modify the above algorithm. Suppose that $\pi(\theta|y_{obs}, z)$ can be calculated up to a constant of proportionality. Generate several values $\theta_l, l = 1, 2, \ldots, L$ from the approximate multivariate normal distribution in (1). Calculate the importance ratios, $w_l = \pi(\theta_l|y_{obs}, z)/\phi(\theta_l|\widehat{\theta}, -\widehat{V}^{-1})$ where $\phi(x|\mu, \Sigma)$ is the multivariate normal density function with mean μ and the covariance matrix Σ, evaluated at x. Resample θ^* from $\theta_l, l = 1, 2, \ldots, L$ with

probability proportional to w_l. The following steps describe the implementation:

(a) Normalize the importance ratios so that they add up to 1. That is, define $w_l^* = w_l / \sum_l^L w_l$.

(b) Form cumulative sums, $c_i = \sum_l^i w_l^*, i = 1, 2, \ldots, L - 1$, and define $c_o = 0$ and $c_L = 1$.

(c) Generate a uniform random number, u, and define $\theta^* = \theta_i$ if $c_{i-1} < u \le c_i$.

The above procedure is called the sampling-importance-resampling (SIR) algorithm for generating values from $\pi(\theta | y_{obs}, z)$. Instead of generating from a normal distribution, sometimes it is useful to generate from a longer tail distribution such as multivariate t with 5 to 10 degrees of freedom with location parameter $\widehat{\theta}$ and scale matrix $-\widehat{V}^{-1}$ (note that the denominator in the importance ratio is the density from which the values of θ have been drawn). Approaches discussed in (1) and (2) could be modified by drawing values in a transformed scale $\psi = h(\theta)$ (where the normal approximation may be better) and then retransform to the original scale $\theta = h^{-1}(\psi)$. For example, when θ is a proportion the transformations, $\text{logit}(\theta) = \log[\theta/(1 - \theta)]$ or $\sin^{-1} \sqrt{\theta}$ are better suited for normal approximation.

3. Suppose $\widehat{\theta}$ is an estimate of θ (such as posterior mode) that can be easily obtained, but the second derivative of the log posterior is not easy to obtain and, again, suppose that $\pi(\theta | y_{obs}, z)$ is easy to calculate up to a constant of proportionality. The following is an approximation of the SIR algorithm given in (2):

(a) Generate several values of complete data $y_i^{(l)}$ from $f(y_i | z_i, \widehat{\theta})$ and retain $D_l = \{y_{i,obs}^{(l)}, z_i, i = 1, 2, \ldots, n; l = 1, 2, \ldots, M\}$ and discard all the generated values of missing observations.

(b) Calculate the estimate $\theta_l, l = 1, 2, \ldots, M$ using D_l.

(c) Calculate $w_l = \pi(\theta_l | y_{obs}, z)$.

(d) Resample a value, θ^*, from $\{\theta_l, l = 1, 2, \ldots, M\}$ with probability proportional to w_l.

(e) Generate imputations from $f(y_{i,mis} | y_{i,obs}, z_i, \theta^*)$.

4. **Gibbs sampling.** Often it is easier to draw values from the complete data posterior density of θ, $\pi(\theta|y_{obs}, y_{mis}, z)$ where $y_{mis} = \{y_{i,mis}, i = 1, 2, \ldots, n\}$. The Gibbs sampling is an iterative strategy: At iteration t, draw a value of θ, $\theta^{(t)}$, from $\pi(\theta|y_{obs}, y_{mis}^{(t-1)}, z)$ and then draw $y_{i,mis}^{(t)}$ from $f(y_{i,mis}|y_{i,obs}, z_i, \theta^{(t)})$. This iteration is continued until the effect of starting value is eliminated.

The field of Bayesian computations continues to evolve, as new approaches for drawing values from the posterior distribution are being developed and implemented. Except for some simple models (i.e., multivariate normal, log-linear model, etc.) software is not available in the missing data context. Implementation of these methods is still a challenge and requires considerable programming efforts.

3.4 Example

A case-control study was conducted to assess the relationship between dietary intake of omega-3 fatty acids and primary cardiac arrest (PCA) (defined as a sudden pulseless condition in the absence of any prior history of heart disease). In particular, the two fatty acids docosahexaenoic acid (DHA) and eicosapentaenoic (EPA) are of interest as they are not synthesized by the body and mostly derived through dietary intake of fish. The two omega-3 fatty acids were measured in the red-cell membrane and summed (REDOMEGA3). This measure is relative to all other measurable fatty acids in the membrane. PCA is a binary outcome variable, 1 for cases (disease) and 0 for controls (no disease). The primary model of interest is the logistic regression model,

$$\text{logit} Pr(\text{SCA} = 1|\text{REDOMEGA3}) = \beta_o + \beta_1 \text{REDOMEGA3}.$$

Unfortunately, REDOMEGA3 is missing for several cases and controls. As is typically the case in many practical applications, several covariates are available that are predictive of the variable with missing values. The study, for a different purpose, collected dietary intake of various types of fish using a food frequency questionnaire. The food frequency questionnaire also elicited information about serving size, which allowed computing an "estimate" of

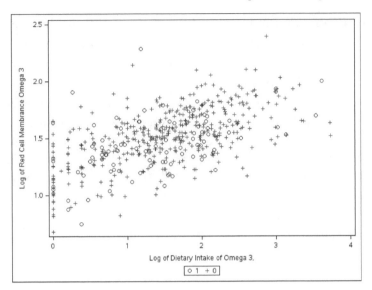

Figure 3.1: Scatter plot of red-cell and dietary values of omega-3 fatty acids on log-scale

omega-3 fatty acid intake. A glitch was that many cases did not survive and the dietary information had to be obtained from the spouse.

The variable DIETOMEGA3 is an estimate from all spouses of cases and controls reporting about their spouses dietary intake. This measure is in absolute intake of DHA+EPA and is not directly comparable to REDOMEGA3. Nevertheless, DIETOMEGA3 may provide information about the missing REDOMEGA3. A few cases and controls with missing values in DIETOMEGA3 are ignored in this analysis.

Figure 3.1 provides a scatter plot of log(REDOMEGA3) and log(DIETOMEGA3 + 1) (1 was added to avoid taking the logarithm of 0's as there were subjects who reported not taking fish at all or insignificant amount to estimate the intake). Given that this is a case-control study, separate indicators for cases ("+") and controls ("o") are used in the scatter plot. The scatter plot shows a linear relationship for both cases and controls and the correlation coefficient between them is about 0.6.

Table 3.1 gives the mean and standard deviation for four subgroups, case/control and whether or not missing red-cell membrane values. The mean and standard deviation of dietary omega 3 for those with and without missing red cell values are similar for both cases and controls. In a complete-case analysis

Table 3.1: Mean (SD) of red-cell membrane and dietary intake of omega-3 fatty acids

Status	Missing Red Cell Value?	Sample Size	Red Cell Omega-3	Dietary Omega-3
Case	Yes	252	—	4.23 (5.87)
	No	95	4.31 (1.16)	4.51 (6.16)
Control	Yes	148	—	4.98 (5.46)
	No	403	4.73 (1.16)	5.41 (5.60)

only 95 cases and 403 controls will be used and 252 cases (73%) and 148 (27%) controls will be discarded. The overall missing data percentage is 44.5%.

The complete-case logistic regression analysis with PCA as the dependent variable and REDOMEGA3 as the predictor results in the estimated regression coefficient of -0.3444 with the standard error of 0.1098. The question is how much information can be recovered through imputation using a reasonable proxy measure, the dietary intake of omega-3 fatty acids.

Imputation of missing values is carried out separately for cases and controls. This will maintain the differences in the observed data distributions between the two groups in the imputation process. Not using separate models will be equivalent to imputing under the null (no difference) which is not appropriate. For controls, a regression analysis of REDOMEGA3 on DIETOMEGA3 results in the estimated regression coefficient as $\widehat{\beta}_o = (1.24568, 0.18039)^t$, residual variance as 0.03999 on 401 degrees of freedom and, for the sake of completeness, the Cholesky decomposition of the inverse of the cross-product matrix as

$$T_o = \begin{bmatrix} 0.10668471 & 0 \\ -0.05419328 & 0.02861489 \end{bmatrix}$$

The following are the steps for generating the imputation of $\log(REDO\text{-}MEGA3)$ for the controls:

1. Generate a chi-square random variable, u, with 401 degrees of freedom. Since the sample size is large, one can approximate $u = 401 + z_1\sqrt{2 \times 401}$ where z_1 is a random normal deviate. To be more precise, one can generate 401 random normal deviates and construct u as the sum of squares of these deviates. Define $\sigma_o^{*2} = (401 \times 0.03999)/u$.

2. Generate two random normal deviates, z_2 and z_3, and define

$$\beta_o^* = \begin{bmatrix} \beta_{oo}^* \\ \beta_{o1}^* \end{bmatrix} = \begin{bmatrix} 1.24568 \\ 0.18039 \end{bmatrix} + \sigma_o^* \begin{bmatrix} 0.10668471z_2 \\ -0.05419328z_2 + 0.02861489z_3 \end{bmatrix}.$$

3. For the control subject j with missing value, generate a random normal deviate, z_j to construct

$$\log(REDOMEGA3_j*) = \beta_{oo}^* + \beta_{o1}^* \log(DIETOMEGA3_j+1) + \sigma_o^* z_j.$$

Similarly for the cases, the estimated regression coefficient is $\widehat{\beta}_1 = (1.22511, 0.15199)^t$, the residual variance 0.04856 with 93 degrees of freedom and the Cholesky decomposition of the inverse of the cross-product matrix,

$$\begin{bmatrix} 0.1967918 & 0 \\ -0.1074688 & 0.06565849 \end{bmatrix}.$$

The following two scatter plots in Figure 3.2 provide the observed and imputed values $log(REDOMEGA3)$ plotted against the $log(DIETOMEGA3 + 1)$ both for cases and controls. Imputed values are indicated by "o" and observed values with "+." For both cases and controls, the imputed and observed values exhibit similar properties, and thus generate a plausible completed-data set from the original population of cases and controls.

Table 3.2 gives the sample size, mean and standard deviation of the observed and imputed values. After imputation, the imputed variables are re-transformed to the original scale and the same logistic regression model was

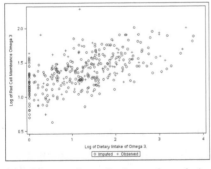
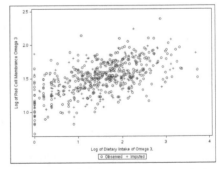

(a) Scatter plot of observed and imputed log(REDOMEGA3) versus observed log(DIETOMEGA3+1) for the cases

(b) Scatter plot of observed and imputed log(REDOMEGA3) versus observed log(DIETOMEGA3+1) for the controls

Figure 3.2: Comparison of observed and imputed values

Table 3.2: Summary statistics of the observed and imputed logarithm of the red-cell membrane omega-3 fatty acids by case-control status

Sample	Status	Sample size	Mean	SD
Case	Observed	95	1.428	0.252
	Imputed	252	1.372	0.259
	Combined	347	1.387	0.258
Control	Observed	403	1.523	0.248
	Imputed	148	1.500	0.248
	Combined	551	1.517	0.248

fit, resulting in the estimated regression coefficient as -0.4579, which is much stronger than the complete-case estimate, -0.3444. The imputed data standard error is meaningless as it assumes that the imputed values are real and, hence, not provided. Also, with the amount of missing data, the difference between the complete-case and imputed estimates may be subject to considerable variation.

This example uses information from an auxiliary variable to impute missing the values in the key variable of interest. In this case, the auxiliary variable is potentially related to the outcome (PCA) as well. In practice there could be many such variables for each variable with missing values. The advantage of the imputation approach is the ability to leverage all such variables in order to recover information lost due to missing values. It is possible that these auxiliary variables may have missing values as well. Methods to handle missing values in many variables and of many types are needed, and the next section describes a generalization of the basic regression method discussed in this section.

3.5 Sequential Regression Imputation

In practice, a data set may have several variables with missing values and differing types such as continuous, categorical, count, etc. Usually, there are structural dependencies between variables in a data set. For example, suppose that the question is asked of a current smoker "How long have you been smoking?" and the person refused to answer, it is known that the years smoked has to be less than age, at the least. Now suppose that the question was asked, "Did you smoke as a teenager?" and the response is no. This gives a narrower range for years smoked. The data sets to be analyzed in practice may contain many such variables.

Sometimes the question is asked only of a subset and therefore any "not applicable" variables are not to be treated as missing data. For example, if the person was asked "Have you ever smoked?" and the answer is no, then all subsequent questions related to smoking are not relevant for this subject.

The variables may also have to be imputed in a certain order. For example, the subject was asked "Q1: Do you plan to vote in the coming election?" with the response option (1) yes, (2) no and (3) don't know. The follow-up question is "Q2: Do you support the ballot initiative A?" with the response option (1) yes, (2) no, (3) undecided. If the goal is to predict the success of the ballot initiative A, then first Q1 has to be imputed and then Q2 for the subset with observed or imputed Q1 to be "yes".

Developing explicit models (joint distributions) or performing hot deck imputations are difficult, if not impossible. A pragmatic approach is to consider a variable by variable imputation but using all the relevant information as predictors. A sequential regression or chained equations approach is one such pragmatic approach. Various other names have been given to this approach such as fully conditional specification or flexible conditional models, etc.

Suppose that U is a collection of variables with no missing values and $Y_1, Y_2, \ldots, Y_p$ are p variables with missing values. Though it is not necessary, suppose that the variables are ordered by number of missing values from lowest for Y_1 and largest for Y_p. An alternative approach is to order on the basis of dependence on other variables from "least dependent" to "most dependent." As discussed later, the ordering will have no effects as the imputed values on any variable will eventually depend on all other variables. Furthermore, the pattern of missing data is assumed to be arbitrary.

The sequential regression approach is an iterative procedure. In the first iteration, Y_1 is regressed on U and the missing values are imputed. An explicit regression model, a hot deck or predictive mean matching may be used to create imputed values. Let $Y_1^{(1)}$ denote the filled-in version of Y_1. Now Y_2 is imputed using $(U, Y_1^{(1)})$ as covariates. Let $Y_2^{(1)}$ denote the filled-in version of Y_2. This process continues until Y_p is imputed using $(U, Y_1^{(1)}, Y_2^{(1)}, \ldots, Y_{p-1}^{(1)})$.

We cannot stop at iteration 1 because imputation of Y_1, for example, fails to exploit the observed information from $Y_2, Y_3, \ldots, Y_p$. Iterations $t = 2, 3, \ldots$, proceed in the same manner except that all other variables (with some filled at the current and the rest in the previous iterations) are used in imputing each variable. Specifically, at iteration 2, Y_1 is reimputed using $U, Y_2^{(1)}, Y_3^{(1)}, \ldots, Y_p^{(1)}$ as predictors; Y_2 is reimputed using $U, Y_1^{(2)}, Y_3^{(1)}, \ldots,$ $Y_p^{(1)}$ as predictors, etc. In general, at iteration t, Y_j is reimputed using

$$U, Y_1^{(t)}, Y_2^{(t)}, \ldots, Y_{j-1}^{(t)}, Y_{j+1}^{(t-1)}, \ldots, Y_p^{(t-1)},$$

as predictors. The iteration is continued for a few times in order to fully use the predictive power of the rest of the variables when imputing each variable. Empirical analysis show that 5 to 10 iterations are sufficient to condition the imputed values on any variable on all other variables.

3.5.1 Details

This section provides a detailed description using a parametric framework. Let Y_o be the observed vector for the variable to be imputed and X be the full sample predictor matrix based on all other variables which is also partitioned into X_o and X_1 where X_o corresponds to subjects in Y_o and X_1 corresponds to subjects with missing values in Y. For the clarity of presentation, suppress the notation for iteration.

1. For a continuous variable, fit a linear regression model of Y_o on X_o and obtain estimated regression coefficient $\widehat{\beta}$, the residual variance $\widehat{\sigma}^2$ and the inverse of the cross-product matrix $(X_o^t X_o)^{-1}$. Let T be the Cholesky decomposition of $(X_o^t X_o)^{-1}$ such that $TT^t = (X_o^t X_o)^{-1}$. Let m be the sample size in this analysis and p be the number of regression coefficients. The imputations are carried out as follows:

   (a) Generate a chi-square random variable, u, with $m - p$ degrees of freedom and define $\sigma_*^2 = (m - p)\widehat{\sigma}^2/u$.

   (b) Generate a vector, z, of p standard normal deviates and define, $\beta_* = \widehat{\beta} + \sigma_* T z$.

   (c) Generate a vector, v, of $n - m$ standard normal deviates and define the imputed values as $Y_{*1} = X_1 \beta_* + \sigma_* v$ where n is the sample size, where X_1 is the predictor matrix corresponding to nonrespondents.

X_o (and X_1) is determined using the standard regression diagnostics, such as scatter plots, residual plots and other graphical techniques to develop a good fitting model. The dependent variable may be transformed (like in the example given in the previous section) to achieve normality of the residuals. The variables may be either transformed back at the end of all the iterations or before proceeding to the next variable. Instead of parametric regression model, hot deck imputation based on the propensity score and the predicted values of Y can be used. The goal is to make sure that the model provides a good basis for predicting the missing values.

2. If Y is binary then a logistic regression model may be used. The following are the steps to generate the imputed values:

   (a) Fit a logistic regression model with Y_o as the dependent variable and X_o as independent variables. Let $\widehat{\beta}$ be the maximum likelihood estimate of the regression coefficient and let $\widehat{V}$ be the estimated variance-covariance matrix.

   (b) As before let T be the Cholesky decomposition of $\widehat{V}$ such that $TT^t = \widehat{V}$. Generate a vector, z, of p standard normal deviates and define $\beta_* = \widehat{\beta} + Tz$. If the normal approximation is not reasonable (which can be checked by plotting the likelihood function against the parameters) then SIR algorithm could be used to generate β_*.

   (c) Using β_* compute the predicted probability p_{*1} for the nonrespondents. That is, define the linear predictor $L_{*1} = X_1\beta_*$ and then define $p_{*1} = 1/(1 + \exp(-L_{*1}))$.

   (d) Generate a vector, u, of $n - m$ uniform random numbers between 0 and 1. Set Y_{*1} to 1 or 0 depending upon whether the generated value of u is less than or greater than equal to the corresponding predicted probability in the vector p_{*1}.

Hoshmer-Lemeshaw goodness of fit tests can be used to develop a good prediction model. This is similar to the techniques used while developing a response propensity model. Instead of using parametric model, hot deck with imputation cells based on the response propensity and predicted probability can be used.

3. For a count variable, a Poisson regression model may be used to develop imputation. The Poisson regression model specifies $Y_o \sim Poisson(n_o\lambda_o)$ where $\log(n_o) + \log\lambda_o = X_o\beta$ or $\log\lambda_o = -\log n_o + X_o\beta$. That is $-\log n_o$ is the offset such as person-years of follow-up or any other suitable denominator.

4. For a nominal variable, multinomial logit model may be used. Suppose that Y can take k levels. The model is

$$\log Pr(Y = j|X) = X\beta_j$$

with $\sum_j Pr(Y = j|X) = 1$. As in the logistic model, obtain the maximum likelihood estimate of the regression coefficients, its covariance matrix and perturbation β_{j*}. Next, compute the predicted values for each nonresponding subject, $p_{j*}, j = 1, 2, \ldots, k$. Generate a uniform random number, u and impute level j if $\sum_l^{j-1} p_{l*} < u \le \sum_l^j p_{l*}$.

5. Mixed or semi-continuous variables occur frequently in many practical applications. For example, real estate income. Most people may have 0 as the income value (no real estate) and a continuous value for the rest. This type of variable can be handled using a two part model. First, impute 0 or non-zero using the logistic regression model and then, conditional on being non-zero use the linear regression model (that is, subset the data for subjects with those observed or imputed as having real estate income) to impute continuous values.

6. For an ordinal variable with k levels, the following proportional odds model may be used:

$$\pi_j = Pr(Y \le j) = \frac{\exp(\alpha_j + X\beta)}{1 + \exp(\alpha_j + X\beta)}$$

with $Pr(Y = j) = \pi_j - \pi_{j-1}$ used to impute the level.

3.5.2 Handling Restrictions

Consider now some restrictions on the imputed values. Suppose that for a continuous variable, the imputed value for the subject i should be between a_i and b_i. The imputed draws are then made from a truncated normal distribution. Similarly, for a categorical variable with k levels, $1, 2, \ldots, k$, if for the subject i, the imputed values can be either 1 or 2 then adopt the following procedure. Let p_{ij*} be the predicted probability for subject i to belong to level j. Compute $\pi_{i1*} = p_{i1*}/(p_{i1*} + p_{i2*})$ and $\pi_{i2*} = 1 - \pi_{i2*}$. Generate a uniform random number and define the imputed value to be 1, if the uniform number is less than or equal to π_{i1*} and 2 otherwise.

This strategy will also be useful in handling missing data in survival or time-to-event analysis. Suppose that X is a set of covariates with some missing values, Y is time-to-event and C is the censoring indicator. Define a variable T which is equal to Y for actual time-to-event, and set T to missing if the observation is censored. The imputation of T has to be greater than or equal to the corresponding Y for the censored observations. Some transformation of survival times will be needed to achieve normality of the residuals.

Sometimes a particular variable is applicable only for a subset of cases. Suppose that a question asked is "Q1: Have you ever smoked cigarettes?" with yes/no response options. A follow-up question is "Q2: Do you smoke cigarettes now?" with yes/no response options. Q2 is applicable to subjects who responded yes to Q1. Imputation of Q2 has to be restricted to those who responded yes to Q1. Thus, the missing values in Q1 has to be imputed first, and then impute missing values in Q2. Obviously if Q1 is missing then Q2 is also missing.

There are two possible ways to handle this situation. Create a new variable by combining Q1 and Q2, into a three category variable with coding 1: never smoker (Q1=no), 2: former smoker (Q1=yes, Q2=no) and 3: current smoker (Q1=yes , Q2=yes). Set all subjects with missing Q1 or Q2 to missing for the newly created variable. The restriction is imposed as discussed above by imputing levels 2 or 3 if Q1=yes.

The second option is to impute the missing value in Q1 first by using a logistic regression model, then subset the data with subjects observed or imputed Q1=yes, and then use another logistic regression model to impute the missing values in Q2. This strategy maintains the question structure and gives more flexibility to the analyst.

Now that Q1 and Q2 have been imputed, the question arises, "How to use Q1 and Q2 as predictors in the imputation of the next variable, say, blood pressure"? One strategy is to recode Q1 and Q2 to create a new three category variable with never, former and current smoker as categories and use two dummy variables as predictors. This requires recoding of variables before each regression which may be inconvenient. The second option is to treat Q2 with three categories yes, no and not applicable. Use one dummy variable for Q1 and two dummy variables for Q2 (treating not applicable as the reference category). Though we are using more dummy variables than necessary in the prediction equation, the advantage of this approach is that it avoids recoding and facilitates automating the imputation procedure. Also, the goal in the imputation is to get a good prediction equation that uses all the variables. The model need not be parsimonious and not purely viewed through a substantive research perspective (for example, the individual regression coefficients may not be interpretable, but a combination of them may be interpretable).

One of the variables of interest may be a scale that is constructed by summing, say, k items (which may be binary 0/1 or 3 to 5 point Likert type item). Consider two situations. In the first situation, subjects responded either to all the items or none of the items and in the second, responses from subjects are mixed. Some provided all, some provided none and the rest provided some and not the others. In the first situation, the scale may be directly imputed (perhaps bounded by the minimum and maximum value possible).

In the second situation, it is better to impute the individual items and then construct the scale by summing after the imputation. This strategy uses partial information and reflects both intra-subject and inter-subject variation in the imputation process. This strategy may not be feasible when the number items and the scales are large. An intermediate solution is to impute the scale, but use the sum of the observed responses as the lower bound and the maximum possible value as the upper bound.

3.5.3 Model Fitting Issues

The advantage of the sequential regression approach is that it reduces the problem of imputation modeling to finding a sequence of good fitting regression models using all available information. All exploratory data analysis techniques can be used to find these regression models. Developing a good fitting model is an iterative process. First, develop a working model based on exploratory data analysis and substantive understanding, check the residuals and perform other model diagnostics, refine the model, if necessary and again perform all the checks, refine the model, etc. Some of these steps can be automated (like generating plots and outputs from model diagnostics).

Sometimes the number of variables may be large making such model building task difficult and time consuming. One potential way to save on model building tasks is to reduce the covariate space for each regression model using a principal component analysis (PCA) of the covariates. Suppose that Y_j is the variable being imputed and $X = (U, Y_1, \ldots, Y_{j-1}, Y_{j+1}, \ldots, Y_p)$ are the predictors. Create principle components $P_1, P_2, \ldots, P_k$ where k is the number of columns in X. Since the principle components are orthogonal to each other, the problem reduces to finding the best fitting k univariate regression models. This may be automated to some extent. One can drop a few principal components corresponding to small percentage of variance explained. This strategy works for all regression models and greatly simplifies the imputation task.

When the number of covariates is large, the covariance matrix of the parameter estimates may be less stable, and the perturbations β_* may be far from the center of the distribution. This can be a significant problem when the predictors in the models are highly correlated. Here also PCA can be helpful. Another option is to use a ridge regression which amounts to shrinking each regression coefficient towards 0. Many software packages are available to fit these models and provide the needed output to carry out the imputation task: Estimated regression coefficients and their covariance matrix for the specified regression model.

Transformation of the variables, or rescaling, is a powerful tool in model building strategy. The Box-Cox transformation of the dependent variable is a powerful method to achieve normality of the residuals. Suppose that Y_j is a continuous variable to be imputed, and X is the covariate matrix. A Box-Cox power transformation fits the regression model,

$$\frac{Y_j^\lambda - 1}{\lambda} = X\beta + \epsilon$$

where the transformation is $\log(Y_j)$ when $\lambda = 0$. When there are some negative values in Y_j, add a constant (can be empirically determined while choosing λ) to make all of them positive to avoid problems with taking logarithms or odd number power transformation. The parameter λ can be estimated from data or determined by trial and error. There are software packages that will provide an estimate of λ using maximum likelihood. Typically, the variables, once transformed, are used in that scale throughout the imputation process and then retransformed to the original scale after imputations of all the variables and all the iterations are completed.

One has to be careful when transforming and retransforming the variables. For example, when using the logarithmic transformation, a large nonsensical value may be obtained by exponentiating what seemingly appears to be a reasonable value on the logarithmic scale. This is especially problematic when log transformation is used for variables such as income, assets or wealth. A cube-root transformation might be better for such variables. However, a hot deck approach using the response propensity and predicted value might be a more reasonable approach for variables that are difficult to model.

When a regression diagnostics shows heteroscedasticity, it can be incorporated in the modeling process as follows. As before, let Y be the continuous variable being imputed and suppose that the residual plot against a particular predictor x_1 shows evidence of increasing or decreasing variance. The model may be modified as $Y \sim N(X\beta, \sigma^2 x_1^\alpha)$. Suppose $\widehat{\beta}$ is the ordinary least square estimate, and e is the residual $Y - \widehat{Y}$. A simple approach to estimate α is to regress $\log(e^2)$ on $\log(x_1)$, and the slope provides an estimate of α, $\widehat{\alpha}$. Now impute the missing values using the refined regression model, $Y^* = Y/x_1^{\widehat{\alpha}/2} \sim N(X\beta/x_1^{\widehat{\alpha}/2}, \sigma^2)$. After the imputation has been carried out for Y^*, multiply by $x_1^{\widehat{\alpha}/2}$ to get the imputation on the original scale, Y.

3.6 Bibliographic Note

Imputation, just like weighting, may have originated in 1930s or 1940s in many government agencies. The hot deck method was perhaps developed in the U. S. Census Bureau or some other official statistical agency. Ford (1983) provides an overview of the hot deck procedure. A more recent review is Andridge and Little (2010). Single imputation is still prevalent in many statistical agencies despite the problems associated with it. Mean imputation is used less. Regression imputation (imputing the predicted values from a regression model) may be more common and can be appropriate in some cases. Many of these imputation techniques were developed when computational power was limited. For more technical details about hot deck method, regression imputation method and some other variance estimation techniques see Kim and Shao (2014).

The sequential regression approach was first proposed by Kennickell (1991) for continuous variables in the survey of consumer finances. Brand (1999) in a doctoral dissertation develops this methodology further (names it variable-by-variable imputation). Van Buuren and Oudshoorn (1999) in a technical report introduced a version of the method and called it multivariate imputation by chained equations and introduced a software MICE. Raghunathan et al (2001) developed the methodology for several types of variables, incorporated bounds and restrictions and termed it, sequential regression multivariate imputation (SRMI). A SAS based and stand-alone software for implementing this procedure is available via web site (www.iveware.org). Another excellent source for software for performing imputation is (www.multiple-imputation.com). The sequential regression approach has been implemented in STATA, another popular statistical software. See Royston (2004). For using R package implementation see van Buuren (2012).

3.7   Exercises

1. **Project**. For this project use the NHANES data downloaded for the exercise in Chapter 2. Perform a hot deck imputation of the missing values using the methods discussed in this chapter. Compare the hot deck and weighted estimators.

2. Compare the sampling variance of the single imputation hot deck estimator (treating the imputed values as real) and compare it with the sampling variance of the weighted estimator. Discuss the extent of under estimation of the sampling variance due to ignoring the imputation uncertainty.

3. Perform a regression analysis of Hemoglobin A1c on the covariates and develop an imputation model. Using the available software or by writing your own code, generate imputation of the missing values. Compute the sample mean from the imputed data and its sampling variance (ignoring the imputation uncertainty). Compare the hot deck and model-based imputation estimates and their standard errors to the weighted estimate and its standard error.

4

Multiple Imputation

4.1 Introduction

It has been repeatedly emphasized that the imputed values should not be treated as the real or actual values for the nonrespondents. However, if a singly imputed data set is analyzed using any standard software, then they are treated as real values, therefore, the stated standard errors will understate the uncertainty. The fact that imputations are guesses (or plausible values) created from the observed data should be reflected in the standard error calculations. Multiple imputation is a mechanism for adding this additional uncertainty.

Under this approach, the imputation process is repeated several times. Each set of imputed values, when combined with the observed set of values, results in a completed data set. Each completed data set is analyzed to obtain estimates of the parameters, their completed-data standard errors and test statistics. The variation across the completed data quantities is used to capture the additional uncertainty due to imputation. The completed data estimates, standard errors, test statistics, etc. are combined to form a single inference.

4.2 Basic Combining Rule

Suppose that M imputed data sets have been created. Let $e_1, e_2, \ldots, e_M$ be the estimates of a parameter, θ, and $U_1, U_2, \ldots, U_M$ be the corresponding estimated sampling variances (square of the standard errors). The multiple imputation estimate is defined as the average of the M estimates,

$$\bar{e}_{MI} = \sum_{l}^{M} e_l/M. \tag{4.1}$$

The variance across the M estimates, $B_M = \sum_{l}^{M}(e_l - \bar{e}_{MI})^2/(M-1)$, actually measures the uncertainty due to imputations. The multiple imputation sampling variance is defined as

$$T_M = \bar{U}_M + (1 + 1/M)B_M \tag{4.2}$$

where $\bar{U}_M = \sum_{i}^{M} U_i/M$ is the average of completed-data sampling variances. One can interpret $\bar{U}_M$ as an estimate of the sampling variance that would have been obtained if there had been no missing data and $(1 + 1/M)B_M$ is the additional variance due to missing data. Thus, the ratio

$$r_M = (1 + 1/M)B_M/T_M \tag{4.3}$$

can be viewed as the proportionate increase in the variance due to missing data. This quantity is called the fraction of missing information (FMI). The ratio $a_M = T_M/\bar{U}_M$ is the efficiency of the multiple imputation estimate relative to complete data estimate. Both these quantities are useful to determine the number of imputations needed, the impact of missing data on inferences and to judge the recovery of information from subjects with partial information through imputation. For example, if FMI is smaller than the proportion of subjects that would be discarded in the complete-case analysis then information has been recovered by including the partially observed subjects. This recovery of information depends on the parameter being estimated and the information in the partially observed subjects.

To construct confidence intervals for θ, a t-distribution with degrees of freedom

$$\nu_M = (M-1)/r_M^2 \tag{4.4}$$

is used. Let $t_{\alpha/2,\nu_M}$ be the appropriate percentile of the t distribution with ν_M degrees of freedom, then $100(1 - \alpha)$ % confidence interval for the parameter, θ, is computed as $\bar{e}_{MI} \pm t_{\alpha/2,\nu_M}\sqrt{T_M}$. To test the null hypothesis, $H_o : \theta = \theta_o$, the test statistic is $(\bar{e}_{MI} - \theta_o)/\sqrt{T_M}$ and is referred to a t distribution with ν_M degrees of freedom.

The degrees of freedom, ν_M, is derived assuming that the complete data inference (in the absence any missing values) are based on large samples (infinite degrees of freedom for the complete data analysis). Sometimes, the complete data analysis (that is, without any missing values) has a finite degrees of freedom. For example, the complete data inferences in a regression analysis based on say, $n = 30$ observations with, say, $p = 5$ predictors has $n - p = 25$ degrees of freedom. In such situations, a correction is needed. Suppose ν_{com} is the complete data degrees of freedom. Define,

$$c_M = \frac{\nu_{com} + 1}{\nu_{com} + 3} \nu_{com}(1 - r_M).$$

The revised degrees of freedom is

$$\nu_M^* = \frac{\nu_M c_M}{\nu_M + c_M}. \tag{4.5}$$

Given the t sampling distribution, the sampling variance is actually a multiple of T_M. The variance of a t distribution, with scale σ and degrees of freedom ν, is $\nu\sigma^2/(\nu-2)$. Thus, the standard error of the multiple imputation estimate may also be defined as $\sqrt{\nu_M^* T_M/(\nu_M^* - 2)}$. When ν_M^* is large, this modification may not make a difference.

4.3   Multivariate Hypothesis Testing

The combining rules in the previous section were for a scalar parameter, θ. In many analyses inferential quantity of interest may be a K-dimenstion vector of parameters, θ. Let $e_1, e_2, \ldots, e_M$ be the estimated vectors from the completed data sets with $U_1, U_2, \ldots, U_M$ as the corresponding variance-covariance matrices. Let $\bar{e}_{MI} = \sum_l e_l/M$ be the mean vector of the completed data estimates and $\bar{U}_M = \sum_l U_l/M$ be the average covariance matrix. Let $B_M = \sum_l (e_l - \bar{e}_{MI})(e_l - \bar{e}_{MI})^t/(M - 1)$ be the between imputation variance-covariance matrix. The multiple imputation variance estimate is $T_M = \bar{U}_M + (1 + M^{-1})B_M$. Wald's chi-square statistics for testing the null hypothesis $H_o : \theta = \theta_o$ is $D_M = (\bar{e}_{MI} - \theta_o)^t T_M^{-1}(\bar{e}_{MI} - \theta_o)$.

When M is small relative to k, the matrix B_M may not be of full rank and the inverse of the matrix, T_M, may be unstable. If the effect of missing data on all the parameters are roughly the same, an approximation is $\tilde{T}_M =$

$(1 + g_M)\bar{U}_M$ and the revised test statistic is

$$\tilde{D}_M = (\bar{e}_{MI} - \theta_o)^t \bar{U}_M^{-1} (\bar{e}_{MI} - \theta_o)/(1 + g_M) \tag{4.6}$$

where $g_M = (1 + M^{-1})Tr(B_M \bar{U}_M^{-1})/k$ and tr is the trace of the matrix or the sum of the diagonal elements of the matrix.

An approximate sampling distribution of $\tilde{D}_M/k$ is an F distribution with k and w_M degrees of freedom where

$$w_M = 4 + (t - 4)[1 + (1 - 2t^{-1})g_M^{-1}]^2 \tag{4.7}$$

provided $t = k(M - 1) > 4$. If $t \leq 4$ then define $w_M = t(1 + k)(1 + g_M^{-1})^2/2$.

The derivation of w_m assumes that the complete data analysis is based on large samples (infinite degrees of freedom). When the degrees of freedom for the complete data analysis is ν_{com} then the modification of w_M (similar to equation (4.5)) is

$$w_M^* = \frac{w_M c_M}{w_M + c_M} \tag{4.8}$$

where

$$c_M = \frac{(\nu_{com} + 1)\nu_{com}}{(\nu_{com} + 3)(1 + g_M)}.$$

4.4   Combining Test Statistics

The combining rules, discussed so far, requires estimates and standard errors for the single parameter inference and the estimates and their covariance matrices for the multiparameter inference. There are many situations where the completed data inferences are based on test statistics or the p-values. For example, suppose that a goodness of fit or association test has been performed on each completed data set yielding M chi-square statistics $d_1, d_2, \ldots, d_M$. A combining rule for these test statistics is as follows. Let $\bar{d}_{MI} = \sum_l^M d_l/M$ be the average of the test statistics. Let $a = \sum_l \sqrt{d_l}/M$ and $b = \sum_l (\sqrt{d_l} - a)^2/(M - 1)$. The test statistic is defined as

$$D_d = \frac{\bar{d}_{MI}/k - (M + 1)r_d/(M - 1)}{1 + r_d} \tag{4.9}$$

where $r_d = (1 + M^{-1})b$ and is referred to an F distribution with $\nu_d = k^{-3/M}(M - 1)(1 + r_d^{-2})$ degrees of freedom.

Table 4.1: Maternal smoking and child wheeze status from the six city study

Maternal	Child's Wheeze Status			
Smoking	No Wheeze	Wheeze with Cold	Wheeze without Cold	Missing
None	287	39	38	279
Moderate	18	6	4	27
Heavy	91	22	23	201
Missing	59	18	26	

The procedure given above can be used for testing hypothesis involving a vector of parameters. For example, assessing the significance of regression coefficients for a block of predictors in a regression model. The above formula can also be applied to the likelihood ratio statistics, $d_1, d_2, \ldots, d_M$ (logistic regression model, for example). Chapter 5 provides alternatives that can be applied for the analysis of variance (ANOVA).

The data summarized in Table 4.1 was collected as a part of a six city study and is reported in Lipsitz, Parzen and Molenberghs (1998). Two categorical variables are maternal smoking status (X_1) classified as (none, moderate and heavy) and the child's wheezing status (X_2) (none, wheeze with cold, wheeze without cold). Total sample size is $n = 1138$ with 528 subjects having both variables measured, 507 missing wheeze status (smoking status available) and 103 missing smoking status (wheeze status available).

One of the goals is to assess the strength of association between these two variables. The missing values were imputed using the sequential regression approach: X_1 was imputed using a multinomial logit model with X_2 as a predictor, and X_2 was imputed using a multinomial logit model with X_1 as a predictor. A total of $M = 20$ imputations were created. The sequential regression was run for 50 iterations.

From each completed table, the Pearson chi-square statistic with $k = 4$ degrees of freedom was computed. The average of these 20 chi-square statistics, $\bar{d}_{MI}$, is 34.784. The variance of the square root of the test statistics, b, is 1.681. Thus $r_d = (1 + 1/20) \times 1.681 = 1.765$. The value of the test statistics D_d is

$$D_d = \frac{34.78/4 - (20+1)/(20-1) \times 1.765}{1 + 1.765} = 2.439.$$

This statistic is referred to as an F distribution with the numerator degrees of freedom k and the denominator degrees of freedom, $\nu_d = 4^{-3/20} \times (20 - $

Table 4.2: Cell specific percentages (SE) [FMI%] for the data from the six city study

Maternal Smoking	Child's Wheeze Status		
	No Wheeze	Wheeze with Cold	Wheeze apart from Cold
None	47.42 (1.7)[24.2]	7.19 (1.08)[49.8]	7.33 (1.02)[42.4]
Moderate	3.19 (0.63)[30.6]	1.16 (0.43)[46.0]	0.98 (0.48)[62.9]
Heavy	20.50 (2.03)[65.2]	5.63 (1.09)[61.0]	6.60 (1.26)[66.1]

1) $\times (1 + 1/1.765^2) = 20.387$. For the significance level 0.05, the cut-off or the critical value is 2.856. For the significance level 0.10, the cut-off value is 2.243. Based on this analysis, there is evidence for the association at $\alpha = 0.10$ level but not at $\alpha = 0.05$ level.

The cell probability estimates were also obtained by averaging the 20 completed-data proportions with the standard error computed using the formula for T_M. Table 4.2 summarizes the estimates, their standard errors and the fraction of missing information. The estimates are practically the same as the maximum likelihood estimate given in the original article. A reference for FMI is the fraction of incomplete cases, 54%. For some cells, the complete-case and the MI estimates are quite different and including the partial information from the incomplete cases reduces the bias and also decreases the variance for some cells but not for others.

4.5 Basic Theory of Multiple Imputation

As mentioned earlier, imputations are generally drawn from a predictive distribution of the missing values, and there are many ways to construct it. The combining rules and theoretical framework makes a broad set of assumptions, and one needs to make sure that these assumptions are generally met.

1. The analyst is assumed to use the most appropriate and efficient estimation or inference procedure on complete data (that is, with missing data). Suppose that $\widehat{\theta}_C$ is the complete data estimate and U_C is its complete data sampling variance (square of the complete data standard error). It is assumed that $U_C^{-1/2}(\theta - \theta_C)$ has a normal distribution (approximately) with mean 0 and variance 1.

2. Denote the average of $\widehat{\theta}_C$ over the predictive distribution of the missing values results as $\widehat{\theta}_O$ and its sampling variance with respect to the predictive distribution is assumed to be $E(U_C) + B_O$, where B_O is the increase in variance due to missing values and $E(U_C)$ denotes the expectation over the predictive distribution of the missing values.

3. The imputation procedure is such that when M, the number of imputations tends to infinity, the average $\bar{e}_{MI}$ tends to $\bar{e}_\infty = \widehat{\theta}_O$ and B_M tends to $B_\infty = B_O$.

4. The imputation procedure yields $\bar{U}_M$ to be approximately equal to $E(U_C)$.

5. When the number of imputations, M, is small the following two conditions hold:

   (a) $MB_\infty^{-1}(\bar{e}_\infty - \bar{e}_{MI}) \sim N(0, 1)$.
   (b) $(M-1)B_M/B_\infty \sim \chi^2_{M-1}$.

Under the stated assumptions,

$$\theta | B_\infty, \bar{e}_{MI}, \bar{U}_M \sim N[\bar{e}_{MI}, \bar{U}_M + (1 + M^{-1})B_\infty].$$

The multiple imputation estimate, its variance, t reference distribution and various test statistics are based on the approximation that combines the above equation and 5(b) to create the posterior distribution of θ given the complete data.

4.6 Extended Combining Rules

The asymptotic normality of the distribution of the complete and completed data estimates is a critical assumption in deriving the combining rules. All combining rules are based on some approximation of $Pr(\theta|\mathcal{S})$ where $\mathcal{S} = \{\mathcal{S}_l, l = 1, 2, \ldots, M\}$), $\mathcal{S}_l$ is a set of statistics from the lth completed data set and θ is the parameter of interest. The point estimate $\bar{e}_{MI}$ and the variance T_M approximate, the mean $E(\theta|\mathcal{S})$ and variance, $Var(\theta|\mathcal{S})$, respectively.

There may be situations where the asymptotic normality assumption is not reasonable. For example, let θ be the population proportion and $\widehat{\theta}_l, l = 1, 2, \ldots$ be the sample estimates. The completed data variance is $U_l = \widehat{\theta}_l(1 - \widehat{\theta}_l)/n_l$ where n_l is the sample size. Note that the sample size may vary across the completed data sets, especially for the proportions defined for a subpopulation defined by variables with missing values. The sample size has to be fairly large for the normal approximation, $(\theta - \widehat{\theta}_l)/\sqrt{U_l} \sim N(0, 1)$, to be reasonable.

Two possible options are considered in this section. The first strategy considers a transformation of the parameter to achieve approximate normality. Construct inferences (e.g., confidence intervals) on the transformed scale and then retransform to the original scale. The second strategy uses the mean $E(\theta|\mathcal{S})$ and $var(\theta|\mathcal{S})$ and develops nonnormal approximations for $Pr(\theta|\mathcal{S})$.

4.6.1   Transformation

Consider two possible transformations for inference about a proportion parameter, θ. Let $\phi = \log(\theta/(1 - \theta))$ and $\widehat{\phi}_l = \log(\widehat{\theta}_l/(1 - \widehat{\theta}_l))$. The completed-data variance is, $var(\widehat{\phi}_l) \approx U_l = [n_l\widehat{\theta}_l(1 - \widehat{\theta}_l)]^{-1}$. The confidence interval and significance tests can be calculated on the transformed scale using the standard combining rule described in Section 4.2. If (ϕ_L^*, ϕ_U^*) is the confidence interval for ϕ then the confidence interval for θ is $[\theta_L^* = \exp(\phi_L^*)/(1 + \exp(\phi_L^*))$, $\theta_U^* = \exp(\phi_U^*)/(1 + \exp(\phi_U^*))]$.

An alternative is to define $\phi = \sin^{-1}(\sqrt{\theta})$ for which $U_l = (4n_l)^{-1}$. If (ϕ_L^*, ϕ_U^*) is the confidence interval for ϕ then $[\sin^2(\phi_L^*), \sin^2(\phi_U^*)]$ is the corresponding confidence interval for θ. There are many other transformations of proportions to achieve normality. The logit and arc-sine-square root transformations are most popular and work well in practice.

Inference about the odds ratio in a 2×2 table is another example where it is better to draw inference on the logarithmic scale (since the normal approximation works better on this scale). Let θ be the odds ratio and define $\phi = \log\theta$. In this case $U_l = (1/a_l + 1/b_l + 1/c_l + 1/d_l)$ where a, b, c and d are the four cell sizes. Once the confidence interval for ϕ is constructed then $[\exp(\phi_L^*), \exp(\phi_U^*)]$ is the confidence interval for θ. Testing the null hypothesis $H_o : \theta = 1$ is equivalent to testing $H_o : \phi = 0$.

As a final example, consider a correlation coefficient. Suppose that r_l is the correlation coefficient between variables X and Y based on the completed data set $l = 1, 2, \ldots, M$. Let $z_l = \log[(1 + r_l)/(1 - r_l)]/2$ be the

Fisher's z transformation of the correlation coefficient. The completed-data variance is $1/(n_l - 3)$ where n_l is the completed data sample size. As before, compute $\bar{z}_{MI}$, T_M and ν_M using the the transformed values. If (z_{*L}, z_{*U}) is the multiple imputation confidence interval on the z scale, then $r_{*L} = [\exp(2z_{*L}) - 1]/[\exp(2z_{*L}) + 1]$ and $r_{*U} = [\exp(2z_{*U}) - 1]/[\exp(2z_{*U} + 1]$ forms the confidence interval for the correlation coefficient.

The same strategy may be applied for inferring about the partial correlation coefficient between X and Y given Z, which is defined as the correlation coefficient between the residuals $e_{X.Z}$ from the regression of X on Z and the residuals $e_{Y.Z}$ from the regression of Y on Z. There are only $n_l - 2$ independent residuals (in both $e_{X.Z}$ and $e_{Y.Z}$). Accordingly, the completed data variance is $U_l = 1/(n_l - 5)$. In general, if p variables are included in Z, then $U_l = 1/(n_l - p - 4)$.

4.6.2 Nonnormal Approximation

For the binomial proportion, assuming a noninformative prior, $\pi(\theta) \propto \theta^{-1/2}(1-\theta)^{-1/2}$, the completed data posterior density is

$$\pi(\theta|x_o, n_o, x_l, n_l) \sim Beta(x_o + x_l + 1/2, n_o + n_l - x_o - x_l + 1/2)$$

where (x_o, n_o) and (x_l, n_l) are the observed and imputed values in the completed data posterior distribution of θ. Thus, defining $e_l = (x_o + x_l + 1/2)/(n_o + n_l + 1)$ and $U_l = e_l(1 - e_l)/(n_o + n_l + 2)$, and defining T_M as usual, a beta distribution may be fitted by matching the mean and variance. That is, $Pr(\theta|\mathcal{S}) \approx Beta(a, b)$ where $a/(a+b) = \bar{e}_{MI}$ and $ab(a+b)^{-2}(a+b+1)^{-1} = T_M$. Solving the two equations obtains,

$$a = \bar{e}_{MI} \left(\frac{T_M}{\bar{e}_{MI}(1 - \bar{e}_{MI})} - 1 \right),$$

and

$$b = (1 - \bar{e}_{MI}) \left(\frac{T_M}{\bar{e}_{MI}(1 - \bar{e}_{MI})} - 1 \right).$$

The confidence interval for θ may be constructed using this approximate beta distribution. The R-package (BINOM, function BINOM.BAYES) provides a convenient tool for computing the highest posterior density interval.

Both transformation and nonnormal approximation may be useful to construct inference about the variance parameter (the variance components). Let

σ^2 be the variance parameter, and let s_l^2 be the completed-data estimate based on $n_l - 1$ degrees of freedom. From standard statistical theory,

$$\frac{(n_l - 1)s_l^2}{\sigma^2} \sim \chi^2_{n_l-1}.$$

Defining $\tau^2 = \sigma^{-2}$, the posterior distribution τ^2 is the scaled chi-square distribution $\chi^2_{n_l-1}/[(n_l - 1)s_l^2]$. The completed-data posterior expectation $e_l = 1/s_l^2$ and $U_l = 2/[(n_l - 1)s_l^4]$. Approximate the distribution of τ^2 as scaled chi-square, $a\chi_b^2$, with mean $\bar{e}_{MI}$ and variance T_M. Thus, $ab = \bar{e}_{MI}$ and $2a^2b = T_M$ resulting in

$$a = T_M/(2\bar{e}_{MI}) \, and \, b = 2\bar{e}^2_{MI}/T_M. \tag{4.10}$$

This strategy will be useful while developing analysis of variance based on multiple imputed data sets (See Chapter 5).

4.7   Some Practical Issues

4.7.1   Number of Imputations

How many imputations are needed? The choice of the number of imputations, M, depends on the fraction of missing information. Note that, the degrees of freedom expression in equation 4.4, $\nu_M = (M-1)/r_M^2$ where r_M is the fraction of missing information. For the same degrees of freedom, the parameter with a larger fraction of missing information (relative to a parameter with smaller fraction of information) will need a larger number of imputation. The fraction of missing information varies by the parameter. Generally, ten imputations may suffice if the fraction of missing information is about 20% and five may be sufficient for a smaller fraction of missing information.

The multiple imputation is inherently a simulation technique and thus, the larger the value of M, the more stable the approximation of the posterior distribution of θ. Given the computational power and availability software, generating a large number of imputations is not difficult. Two critical quantities to monitor are $\bar{e}_{MI}$ (defined in equation 4.1) and T_M (defined in equation 4.2). One can be adaptive in choosing M to ensure that these two quantities stabilize.

Many investigators have developed methods for choosing M. A simple rule is to choose $100r$ where r is the largest fraction of missing information. This number may not be available easily. One possibility is to choose $M = 5$, estimate r_M based on this pilot number of imputations, and then increase the number of imputations to $100r_{max}$. Another possibility is to choose the fraction of incomplete observations for a particular analysis as the fraction of missing information. Thus, if the analysis involves discarding 30% of the subjects then choose $M = 30$.

4.7.2 Diagnostics

How to check whether the imputations are reasonable? A general approach is to compare summary statistics (such as mean and standard deviation, skewness, kurtosis, etc.) of the imputed and observed values. These statistics should be similar under missing completely at random and may not be the same if the data are not MCAR. Nevertheless, under modest missing at random assumption, observed and imputed data statistics should be similar.

Now consider a specific approach. Suppose that U is the variable that has some imputed values and $V_1, V_2, \ldots, V_p$ are the predictors used in the imputation model and, for now, assume that they are fully observed. The imputation assumes that the data are missing at random. Suppose that U_{obs} are the observed values, and U^*_{mis} are the imputed values. Under, MAR, if the imputation model is reasonable then

$$f(U_{obs}|V_1, V_2, \ldots, V_p) \equiv f(U^*_{mis}|V_1, V_2, \ldots, V_p).$$

This equivalence can be checked using the response propensity methods as follows. Let R be the response indicator for U, and $p = Pr(R = 1|V_1, V_2, \ldots, V_p)$ be the response propensity which is estimated using, say the best fitting logistic regression model to obtain $\hat{p}$. Using the property of the balancing through propensity score, under the correct imputation model $f(U_{obs}|\hat{p}) \approx f(U^*_{mis}|\hat{p})$.

The balance checking diagnostics that were used in Chapter 2 for the weight construction can be used to asses whether or not, conditional on the propensity score, the observed and imputed values are similar. Specifically, suppose that U^* is a completed data vector with observed and imputed values. Regress U^* on $\hat{p}$ and store the residuals, $\hat{e}$. Construct kernel densities of the residuals separately for the observed and imputed values. If the kernel densities overlap then the imputations are reasonable.

An alternative is to plot U^* versus $\widehat{p}$ and identify the observed and imputed values using different symbols. Under the correct imputation model, the symbols for the observed and imputed values should have be exchangeable across the full spectrum of the propensity score. The scatter plots of the observed and imputed values of log(REDOMEGA3) versus log(DIETOMEGA3 + 1) discussed in Chapter 3 are a simple example of this procedure as there was only one covariate.

What if some covariates $V_1, V_2, \ldots, V_p$ have missing values? Let V_{obs} denote the observed values. Under the correct imputation model, $f(U_{obs}|V_{obs}) \equiv f(U_{mis}^*|V_{obs})$ or, equivalently, $f(U_{obs}|\widehat{p}_{obs}) \approx f(U_{mis}^*|\widehat{p}_{obs})$ where $p_{obs} = Pr(R = 1|V_{obs})$. One way to estimate $\widehat{p}_{obs}$ is as follows. Suppose that the covariates have been multiply imputed along with U, say M times. Let $\widehat{p}_l$ be the propensity score estimate based on l^{th} completed data on $V_1, V_2, \ldots, V_p$. Assuming that the imputations of $V_1, V_2, \ldots, V_p$ are reasonable,

$$\widehat{p}_{obs} = \sum_{l}^{M} \widehat{p}_l/M.$$

The diagnostics proceed as before with using U^* and $\widehat{p}_{obs}$.

4.7.3   To Impute or Not to Impute

Sometimes a question is raised "Q1: I have a lot of missing values in the variable X. Should I impute or not"? For example, suppose that 75% of the observations are missing. Before answering this question, consider the following question "Q2: What is the alternative"? If the answer to Q2 is complete-case analysis, then it does not make sense because one will be discarding 75% of the observations on other variables, an enormous sacrifice to accommodate not imputing the missing values in X. Furthermore, the result may be severely biased, if the missing data mechanism is not MCAR. If the answer to Q2 is to ignore X from the analysis then imputation is no longer necessary, but this answer has altered the scientific question.

As long as the variable is going to be included in the analysis, it is better to deal with missing values through imputation using as much information as possible. The goal is to maintain the original sample in the analysis and do the best one can with the missing values. In this case, the number of imputations have to be large to get stable results.

4.8 Revisiting Examples

4.8.1 Data in Table 1.1

For the data in Table 1.1, cell-specific imputations may be performed using the normal model with cell-specific mean and cell-specific standard deviation. Consider the cell $D = 0, E = 0$ with 299 sample observations (n_{oo}) but only 212 respondents (r_{oo}). The assumed model is $x \sim N(\mu_{oo}, \sigma_{oo}^2)$ with a diffuse prior distribution for the parameters, $\pi(\mu_{oo}\sigma_{oo}) \propto \sigma_{oo}^{-1}$. The sample mean is $\bar{x}_{oo} = -0.57$ and the sample variance $s_{oo}^2 = 0.9^2$. Note from the basic theory that

1. $\sqrt{r_{oo}}(\bar{x}_{oo} - \mu_{oo})/\sigma_{oo} \sim N(0, 1)$

2. $(r_{oo} - 1)s_{oo}^2/\sigma_{oo}^2 \sim \chi_{r_{oo}-1}^2$.

The eighty seven observations can be multiply imputed as follows:

1. Generate a chi-square random variable, u, with $r_{oo} - 1$ degrees of freedom and define $\sigma_{oo}^2* = 211 \times 0.9^2/u$.

2. Generate a random normal deviate z, and define $\mu_{oo}^* = -0.57 + \sigma_{oo}^* z/\sqrt{r_{oo}}$.

3. Generate 87 random normal deviates, $z_1, z_2, \ldots, z_{87}$, and define $x_j^* = \mu_{oo}^* + \sigma_{oo}^* z_j$ as the imputed values.

4. Repeat steps (1)-(3) M times.

The same procedure can be applied for the other three cells. After imputation, fit the logistic regression model with D as the outcome, E and X as the predictors. Table 4.2 provides five imputed data estimates and their standard errors for the regression coefficient of E in the logistic regression model with D as the dependent variable and E and X as predictors. The mean of the regression coefficients for E is $\bar{e}_{MI} = 0.4369$. The average of the squared standard error is $\bar{U}_M = 0.0205$ and the variance of the five estimates is $B_M = 0.0190$ and, thus, $T_M = 0.0205 + (1 + 1/5)0.0190 = 0.0433$ or the multiple imputation standard error is 0.2081. The fraction of missing information is $r_M = (1 + 1/5)0.0190/0.0433 = 52.7\%$ and the degrees of freedom $\nu_M = 4/.527^2 = 14.4$. This leads to a 95% confidence interval as $(.4369 \pm 2.14 \times 0.2081)$.

Table 4.3: Estimates and their standard errors for logistic regression example using the simulated data in Table 1.1

Imputation	E		X	
	Estimate	Standard Error	Estimate	Standard Error
1	0.4119	0.1434	0.5674	0.0794
2	0.4465	0.1435	0.4861	0.0776
3	0.4558	0.1420	0.5299	0.0780
4	0.4487	0.1423	0.5391	0.0798
5	0.4215	0.1441	0.5129	0.0781

4.8.2   Case-Control Study

Revisit the case-control analysis where the missing values in log(REDOME-GA3) was imputed using a model based procedure using the dietary value log(DIETOMEGA3 + 1). The number of missing values is rather large in this example (73% for cases and 27% for controls). Since imputations are being carried out separately for cases and controls, a large number of imputations may be needed to get stable answers. The problem is simple enough to repeat the imputation a large number of times using a variety of software packages.

A total of $M = 100$ imputations were carried out and after each imputation, the variable was retransformed back to the original scale, and then the substantive logistic regression model with PCA as the dependent variable and imputed (or completed) REDOMEGA3 as a predictor was fit. The second column in Table 4.4 provides the estimated regression coefficient, its standard error and the fraction of missing information.

In this example, an auxiliary variable was used to impute the missing values with the hope of recovering information about the missing red cell membrane values. Perform multiple imputation analysis without using the auxiliary variable as a way to assess its usefulness. The last column in Table 4.4 provides the estimate, standard error and the fraction of missing information based

Table 4.4: Estimated regression coefficient, the standard error (SE) and the fraction of missing information (FMI) for the case-control study example based on $M = 100$ imputations

Quantity	Using Dietary Intake	Ignoring Dietary Intake
Estimate	-0.344	-0.330
SE	0.104	0.111
FMI	61.4%	66.3%

Table 4.5: Estimates, standard errors and the fraction of missing information for estimating the mean

Status	Quantity	Controls	Cases
Include	Estimate	4.710	4.278
Dietary	SE	0.056	0.106
intake	FMI	20.5%	67.0%
Exclude	Estimate	4.727	4.313
Dietary	SE	0.059	0.115
intake	FMI	27.9%	73.1%

on $M = 100$ imputations. Including dietary intake in the imputation model reduces the fraction of missing information by about five percentage points (6.4% reduction in the standard error of the regression coefficient).

The efficiency gain in using dietary intake in the imputation process can differ by the type of parameters. Table 4.5 provides information about estimating the mean of red-cell membrane values for cases and controls. Reduction in the standard error and the fraction of missing information are comparable to those given in Table 4.4.

4.9 Example: St. Louis Risk Research Project

Consider data from St. Louis Risk Research Project described in Little and Rubin (2002). One of the objectives is to evaluate the effects of parental psychological disorders on the development of their children. This data involved 69 families classified into one of three categories, $G = 1$ (normal or control families from local community), $G = 2$ (a moderate risk group where one parent was diagnosed to have a secondary schizo-affective or some other psychiatric illness or one parent having a chronic physical illness), $G = 3$ (a high risk group where one parent suffers from schizophrenia or an affective mental disorder). There are 27 families in the normal group, 24 families in the moderate risk group and 18 families in the high risk group. From each family, two children were evaluated for standardized reading (R) and verbal comprehension (V) (both yielding continuous scores) and a binary variable (S) with low or high number of psychological symptoms. The goal is to compare the six variables (reading scores on two children (R_1,R_2), two verbal comprehension course (V_1, V_2) and two binary symptom classification (S_1, S_2)) across the three family categories. Not all variables are available for each family.

A sequential regression approach will be used to multiply impute the missing values in these six variables and then perform a series of analysis comparing the distribution of these six variables across the three groups. A prediction model is needed for each variable using all others as predictors. Since the goal is to compare the three groups, one may not want to pool models across the three groups to avoid biasing the imputations towards the null, i.e., if V_1 were imputed using two dummy variables for G, and using all other variables as predictors. The imputation model assumes that the relationship among the variables is the same across the three groups except for different intercepts. Separate imputation by group may be less efficient as it fails to capitalize on information from one group to impute in others. Here, however, maintaining the group difference is important and thus separate imputations will be carried out (similar to the case-control study example).

In developing models, the first step is to investigate general relationship between these six variables. Figure 4.1 provides scatter plots (pooling data across all three groups) relating V_1 to V_2, R_1 to R_2, V_1 to R_1 and V_2 to R_2. All scatter plots show that, in general, linear regression models are good candidates. Linear regression models were used to impute the missing values in V_1, V_2, R_1 and R_2 and a logistic regression model for S_1 and S_2.

Given a large number of missing observations, a total of 20 imputations were created for each group. Also, it is important to make sure that R_1, R_2, V_1 and V_2 are imputed positive values (normal linear regression with small sample size can have large variance for the predictive distribution, and hence may result in negative values for positive random variables). The imputations were carried out using the sequential regression approach as implemented in IVEWARE, bounds were set for the variables to be positive and the sequence was run for 50 iterations. The imputations could have been carried out using the package MICE in R or MI in STATA.

After imputations, many analysis could be performed to answer the basic question, "Are there differences in the reading and verbal measures and the symptoms across the three groups"? As a preliminary analysis, construct a family level score through averages $R = (R_1 + R_2)/2$, $V = (V_1 + V_2)/2$ and $S = (S_1 + S_2)/2$. The two contrasts of interest for each variable are $E(Y|G=2) - E(Y|G=1)$, $E(Y|G=3) - E(Y|G=1)$ where $Y = R, V, S$.

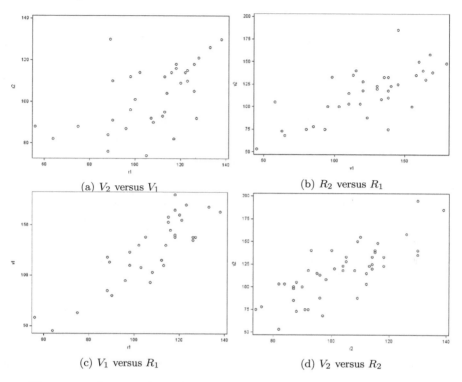

Figure 4.1: Scatter plots of variables in St. Louis risk research project

One may also be interested in a multivariate hypothesis of jointly comparing all three variables between the two groups.

After the imputation, three variables R, V and S were created and the mean and the covariance matrices were computed for these three variables in each group. Table 4.6 provides the means and their standard errors for each variable and group.

Table 4.6: Multiple imputation mean, standard error and the degrees of freedom for three composite variables R, V and S for three groups

Variable	$G = 1$			$G = 2$			$G = 3$		
	Mean	SE	df	Mean	SE	df	Mean	SE	df
R	112.5	2.6	23.8	102.6	2.6	21.1	102.2	5.1	18.2
V	135.5	4.6	19.4	108.9	6.4	17.9	112.8	9.2	15.5
$S(in\%)$	40.4	8.6	14.1	70.6	8.4	14.4	58.2	11.2	15.1

The differences are not statistically significant. For example, the difference between $G = 1$ versus $G = 3$ for R is $112.5 - 102.2 = 10.3$. That is, children in the high risk group score about 10 points lower in the standardized reading. Since the imputations were carried out independently, the standard error of the difference is $\sqrt{2.6^2 + 5.2^2} = 6.1$. The difference in the mean is about 1.7 times the standard error. Similarly, for V, the difference is $135.5 - 112.8 = 22.7$ and its standard error is $\sqrt{4.6^2 + 9.2^2} = 10.3$. The difference in the verbal comprehension score is slightly more than two standard errors lower for the high risk group than the normal group. For the percentage of symptoms, the difference is $40.4.5 - 58.2 = -17.8$ percentage points with the standard error, $\sqrt{8.7^2 + 11.2^2} = 14.2$. The proportion of children having a high number of symptoms in the high risk group is about 1.3 standard error larger than the normal group. Similar analysis comparing $G = 2$ and $G = 3$ shows that these variables have similar mean between these two groups. More formal tests can be performed by formulating these contrasts as regression coefficients in a general linear model or multivariate analysis of variance as discussed in Chapter 6.

Without imputations, data from many families would have been discarded. For example, of the 27 families in $G = 1$ group, only ten families have both R_1 and R_2 measured. That is, a loss of 17 out of 27 observations or 63%. However using imputations, the fraction of missing information is 13.7%. Table 4.7 provides the fraction missing observations (for the composite measures R, V and S and the fraction of missing information by group).

From the table, it is clear that through multiple imputations, considerable information has been "recovered" compared to the complete case analysis based on families with both children providing the data. The recovery of information could be improved by pooling $G = 2$ and $G = 3$ groups and using

Table 4.7: Fraction of missing observations and missing information: St. Louis risk research study

Variable	$G = 1(n = 27)$		$G = 2(n = 24)$		$G = 3(n = 18)$	
	% Missing	FMI (%)	% Missing	FMI (%)	% Missing	FMI (%)
R	63.0	13.7	41.7	21.5	44.4	30.2
V	44.4	26.7	54.2	31.1	44.4	38.6
S	66.7	43.5	50.0	42.2	55.6	40.1

the dummy variable indicating $G = 1$ as a covariate in addition to the six variables R_1, V_1, S_1, R_2, V_2 and S_2.

4.10   Bibliographic Note

Multiple imputation was proposed by Rubin (1978a) and most theoretical and practical considerations are described in the classic book Rubin (1987). The theoretical justification given in this chapter is derived in Rubin (1987) and Raghunathan (1987). Rubin and Schenker (1986) develops the combining rules for a scalar parameter. Multiparameter case is developed in Raghunathan (1987) and refined in Li et al. (1991a). Combining chi-square statistics is also developed in Raghunathan (1987) and extended in Li et al. (1991b).

Barnard and Rubin (1999) developed the modification in equation (4.5). Reiter (2007) further refined the combining rule. Meng and Rubin (1992) developed a method for performing the likelihood ratio test with multiply imputed data sets.

4.11   Exercises

1. The data from the case-control study analyzed in Chapters 3 and 4 is available on the website www.iveware.org as a part of example data sets with IVEWARE package. It is an SAS data set with file name "Test." Extract the variables DHA_EPA, CASECNT and REDTOT. The goal is to fit the logistic model,

    $$\text{logit} Pr(\text{CASECNT} = 1 | \text{REDTOT}) = \beta_o + \beta_1 \log(\text{REDTOT}).$$

    (a) Perform multiple imputation of the missing values using the package of your choice.

    (b) Fit the model given above and construct inferences for β_1.

    (c) Assess the benefit of using DHA_EPA in estimating β_1.

2. Refer to problem (1). Sometimes epidemiologists like to categorize the exposure variable and fit a logistic regression model with the dummy variables for the categories as predictors. Use the quartiles for the controls with no missing data as the cut-off points. Define three dummy variables treating the first quartile as the reference category. Fit the logistic regression model with three dummy variables on both observed and imputed data sets. Compare the results and assess the benefit of using the dietary intake value in the imputation process.

3. Another variable that might be predictive of red-cell membrane values is the fat index. Perform multiple imputation analysis using both the dietary intake and fat index.

4. For St. Louis risk research data (available in Little and Rubin (2002)), perform multiple imputation analysis by combining $G = 2$ and $G = 3$ but include dummy variable for $G = 1$ in the imputation model. Also, perform model diagnostics to assess whether any interaction term between this dummy variable and any of the six substantive variables need to be included in the imputation model.

5. Refer to project 5 in Chapter 1. Perform multiple imputations on each of the 100 data sets, create scatter plots with multiple imputation estimates. Compare these scatter plots with those obtained in project 5 of Chapter 1.

6. To assess the effect of ignoring imputation uncertainty, construct inferences from the first imputed data obtained in problem 5 and compare them with multiply imputed inferences. For each parameter, compute and compare the number of singly and multiply imputed confidence intervals that contain the true value. What are the other ways to compare?

7. The following table provides data from Glasser (1965). The response variable (y) is the logarithm of survival times (in days) for patients with primary lung tumors. The age of the patient (x_1) and performance status rating (x_2) are the two covariates. The starred values are censored observations.

Log Survival Time	Age	Performance Status
1.94	42	4
2.23	67	6
1.94	62	4
1.98	52	6
2.23	57	5
1.59	58	6
2.13	55	6
1.80	63	7
2.32	44	5
1.92	62	7
2.15*	51	7
2.05*	64	10
2.48*	54	8
2.42*	64	3
2.56*	54	9
2.56*	57	9

(a) Perform multiple imputation normal linear regression analysis of the response variable on the two covariates. Treat the response variable for the censored observations as missing and use censoring time as the lower bound.

(b) Given the small sample size, apply the Barnard-Rubin correction while constructing the confidence intervals for the parameters and contrast it with the standard procedure intervals in (a).

(c) This data is also analyzed in Lawless (1982, pages 318-320) using maximum likelihood. Compare multiple imputation and maximum likelihood estimates of the parameters for the same regression model.

(d) For each completed data fit the following regression model,

$$y = \beta_o + \beta_1(x_1 - \bar{x}_1) + \beta_2(x_2 - \bar{x}_2) + \epsilon,$$

where $\epsilon \sim N(0, \sigma^2)$. Construct multiple imputation inferences for the probability of survival beyond eight days for an average age patient with an average performance status.

5

Regression Analysis

5.1 General Observations

Regression analysis, a common technique, is used in many scientific disciplines. These may be linear or nonlinear models or a complex set of models such as the structural equation models, two-stage least squares, censoring, etc. The multiple imputation analysis framework is the same: perform the analysis on each completed data set, extract the meaningful parameter estimates, their covariance matrix, test statistics and combine them using the methods described in Chapter 4. This chapter covers some key issues about imputation and analysis in the context of a regression analysis and provides some additional combining rules. These are specific, to regression analysis such as ANOVA, Partial F-test and R^2 and Adjusted R^2 analysis.

5.1.1 Imputation Issues

Should the outcome variable be imputed in the regression model?

Consider a regression analysis where the missing values are in the outcome variable (y) and the covariates $x = (x_1, x_2, \ldots, x_p)$ are fully observed. Here the complete-case analysis is a valid approach under MAR mechanism for two reasons: First, under MAR, the conditional distribution for respondents, $f(y|x, R_y = 1)$, is the same as for nonrespondents, $f(y|x, R_y = 0)$ where R_y is the response indicator for the variable y and secondly, there is no information about regression in the incomplete cases as y is not observed. Strictly speaking, the imputation of the missing values is adding noise to the "information" content in the observed data about the regression.

In many practical situations a set of good predictors, $z = (z_1, z_2, \ldots, z_q)$, may be available for predicting missing values in the variables (y, x)(outcome or covariates). These additional predictors are not part of the substantive regression model. For example, in the case-control study the dietary intake of omega-3 fatty acid is a predictor of the red cell values but was not part of the substantive or analysis model. It may be better to jointly multiply impute the missing values in (y, x, z) and then perform a subset analysis using the multiply imputed (y, x) without ignoring the imputed values in y. Furthermore, through conditioning on the additional covariates z, the missing data assumption becomes less stringent when compared to ignoring z. This strategy can increase the efficiency depending upon the predictive power of z for (y, x).

A disadvantage of the joint imputation of (y, x, z) is the complexity of model construction, especially, if there are missing values in z as well. The trade-off between increased efforts in imputation modeling and the improvement in the efficiency needs to be carefully investigated before ignoring or using the additional predictors, z. Obviously, no variables in x should be omitted from the regression model as it will lead to biased regression coefficients (omitted variable bias).

There is a subtle issue that needs a discussion. The repeated sampling calculations are based on clearly defining the experiment, the basis for calculating the standard errors and other criteria such as consistency, bias and confidence coverage. Consider the following two different scenarios:

1. A sample is drawn to measure (y, x, z) but the interest centers on estimating
$$f(y|x) = \int f(y|x, z) f(z|x) dz.$$

2. The sample drawn is $\{(y, x), z\}$ and z has no relevance to the relationship between y and x.

In the first scenario, z provides information about the missing y and x and therefore should be used. If the repeated sampling calculations are going to be performed under the scenario 1, ignoring z may lead to loss of efficiency and even bias, especially, if z is related to missing data mechanism.

In the second scenario, z has no information about y or x, even if it is related to missing data mechanism. Under the repeated sampling, z is just a noise and including it in the imputation process leads to loss of efficiency.

In the third scenario, y and z are correlated only through x which may be denoted as $\{(y, x), (x, z)\}$. In this case, the missing values in x and y should be imputed using z but the imputed values in y should be discarded for the regression analysis of y on x but the imputed values in x should be used.

As a practitioner, it is important to use all available information to recover as much information about the missing values as possible. Thus, it requires a careful study of all available variables and their relationship with the variables of interest. Of course, there are some practical constraints and some of the methods were described in Chapter 4 to accommodate large number of variables in the imputation process.

Should the outcome variable be used to impute the missing values in the covariates? Now suppose that the missing values are in y and x_1. The covariates $x_2, \ldots, x_p$ are fully observed. Should one carry out the multiple imputation of missing values jointly for (y, x_1) conditional on $x_2, x_3, \ldots, x_p$? Doing so gives the appearance of a "double dipping," or "endogeniety" where a portion of left-hand-side variables also seems to appear on the right-hand side of the model. The answer to the question may become apparent by considering the complement: Not including y in the imputation of covariates implies that among the imputed values there is no relationship between y and x. This will obviously underestimate the relationship in each completed data set. The general recommendation is, therefore, to condition on all the variables to maintain the plausibility of completed data as a legitimate data from the population. However, the analysis may be carried out using the observed values of y and the corresponding values of x which may include some imputed values in x_1. Obviously, this approach can be generalized to situations where the missing values may occur in many covariates.

To investigate this point further, a simulation study was conducted in the context of imputing income values in the consumer expenditure survey (CEX) conducted by the U.S Bureau of Census for the Bureau of Labor Statistics. The question of interest is whether the reported expenditures (in addition to several other covariates) should be used in the imputation of income.

A population was created by accumulating complete cases from several years of CEX. Generally, this is a better way of creating the population for a simulation study rather than using a statistical model (as in the logistic regression model simulation in Chapter 1). Two hundred independent samples of size 500 each was drawn from this population (before deletion data sets). The data set contained several income and expenditure variables as well as covariates. On each data set, some income values were deleted using an arbitrary missing at random mechanism conditional on the variables in the data set (after-deletion data sets).

Two sets of imputations of missing values in the income variables were carried out: one included expenditure variables as predictors and another excluded them. A log transform was used for the income and a linear regression model was developed to impute the missing values. Imputations were drawn from the corresponding predictive distribution.

Two typical analytical models were chosen for the evaluation. One was a linear regression model with expenditure on food at home as the outcome variable, with income as the primary predictor of interest along with many other covariates. The second model was the left censored regression model or a Tobit model with expenditure on food away from home as the outcome with income and other covariates as predictors. [A Tobit model describes the relationship between a nonnegative dependent variable y and a vector of independent variables x. Suppose that y^* is a latent variable and is related y through the following relationship: $y = y^*$ is $y^* > 0$ and $y = 0$ is $y^* \leq 0$. The latent variable y^* is also related to x through a regression model $y^*|x \sim N(x^t\beta, \sigma^2)$.]

The results are summarized in four scatter plots in Figure 5.1. The values plotted are the estimated regression coefficients for the income variable. The y-axis contains estimates from the actual or before deletion data sets and the x-axis contains estimates from the multiply imputed data sets $(M = 5)$. Figures (a) and (b) are for the OLS model and (c) and (d) are for the Tobit models. Figures (a) and (c) ignore expenditure variables in the imputation process whereas (b) and (d) include expenditure in the imputation model.

It is clear from the scatter plots that excluding expenditures from the imputation model biases the regression coefficients towards zero (that is, the multiply imputed estimates are considerably smaller than the corresponding before deletion data set estimates) for both OLS and Tobit models. When the expenditure variables are included in the imputation model, figures show almost 45 degree relationship between the actual and multiply imputed regression coefficient estimates. The spread in the scatter plot indicates that, in general, the OLS model fits the data better than the Tobit model.

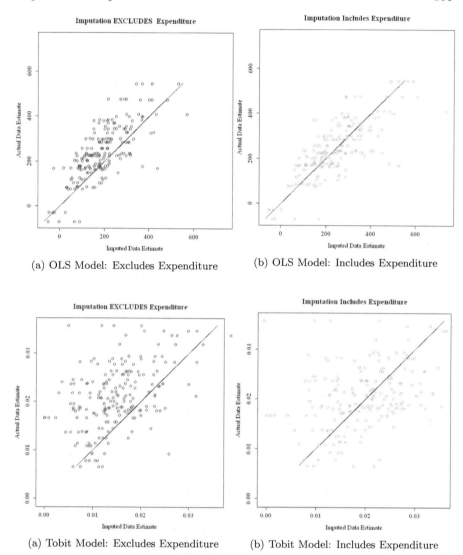

(a) OLS Model: Excludes Expenditure   (b) OLS Model: Includes Expenditure

(a) Tobit Model: Excludes Expenditure   (b) Tobit Model: Includes Expenditure

Figure 5.2: Scatter plots comparing estimates with and without expenditures in the imputation model: CEX simulation study

5.2   Revisiting St. Louis Risk Research Example

A regression model with the reading or verbal scores or high/low symptoms $(R, V$ or $S)$ as dependent variables and the risk group (G) with three categories as the predictor variable may be used to compare the effect of parental psychiatric status on the child development. However, the two measures within each family will be correlated (see scatter plots in Figure 4.1). The correlation between observations within a family could be incorporated using a random effect model. The model development is illustrated for the verbal score which can be applied to the reading score. Model for S will be discussed later.

Let V_{ij} be the verbal comprehension score for child $i = 1, 2$ in family $j = 1, 2, \ldots, 69$. Let $G_{2j} = 1$, if the family is a member of group 2 and 0 otherwise. Similarly, let $G_{3j} = 1$, if the family is a member of group 3 and 0 otherwise. The basic random effect regression model is

$$V_{ij} = \beta_o + \beta_1 G_{2j} + \beta_2 G_{3j} + \gamma_j + \epsilon_{ij}$$

where $b_j \sim N(0, \sigma_F^2)$ and $\epsilon_{ij} \sim N(0, \sigma_E^2)$ all the random effects (bs) and the residuals (ϵs) are all independent of each other. The random effect introduces the correlation between the residuals for the children in the same family and assumed to be the same for all three groups as the variance component σ_F does not depend upon the group. The correlation coefficient between the two observations within the same family is $\sigma_F^2/(\sigma_F^2 + \sigma_E^2)$, called intraclass correlation coefficient. This model can be fit using any standard software package such as SPSS, R, STATA or SAS (PROC MIXED in SAS was used here).

The model was fit on each of the 20 imputed data sets (using PROC MIXED in SAS), point estimates and standard errors for each regression coefficient were combined using the standard combining rule. Table 5.1 summa-

Table 5.1: Multiply imputed random effects model analysis of reading and verbal scores in the St. Louis risk research study

Parameter	Reading Scores				
	Estimate	Lower 95% Limit	Upper 95 % Limit	df	FMI (%)
$G = 1 (\beta_o)$	112.5	106.7	118.1	56.9	11.8
$G = 2$ vs $G = 1$ (β_1)	-9.9	-18.2	-1.5	55.7	12.9
$G = 3$ vs $G = 1$ (β_2)	-10.2	-20.6	0.2	36.1	33.3
	Verbal Scores				
$G = 1 (\beta_o)$	135.5	124.6	146.4	49.4	19.2
$G = 2$ vs $G = 1$ (β_1)	-26.6	-43.6	-9.6	40.4	28.5
$G = 3$ vs $G = 1$ (β_2)	-22.7	-43.8	-1.6	27.7	44.3

Table 5.2: Multiply imputed generalized estimating equation analysis of symptoms in the St. Louis risk research study

Parameter	Symptoms (reporting high)				
	Estimate	Lower 95% Limit	Upper 95 % Limit	df	FMI (%)
$G = 1$ (β_o)	-0.395	-1.142	0.352	28.9	42.6
$G = 2$ vs $G = 1$ (β_1)	1.286	0.198	2.373	32.7	37.5
$G = 3$ vs $G = 1$ (β_2)	0.733	-0.426	1.892	30.4	40.6

rizes the results for β_o, β_1 and β_2 for both outcomes R and V. Both reading and verbal scores are considerably lower for both $G = 2$ and $G = 3$ groups when compared to $G = 1$. Of the four contrasts only one is not significant at the 5% level. The p-value for that contrast is 0.055.

For the binary variable S, the generalized estimating equation framework can be used to accommodate the correlation between observations in the same family. Define $S_{ij} = 1$, if the child i in family j is classified to have high symptoms and 0 otherwise. A logistic model for S_{ij} was fit with G_{2j} and G_{3j} as predictors using PROC GENMOD in SAS. The point estimates and the standard errors were combined using the standard combining rules. Table 5.2 summarizes the results in the same format as Table 5.1. Both tables show that children in the moderate and high groups are worse off than the normal group. In fact, the moderate group may be slightly worse than the high risk group.

5.3 Analysis of Variance

The analysis of variance is a useful technique in many disciplines and is used in both randomized and observational studies. First consider a typical analysis using ANOVA with complete-data, and then consider the same analysis with missing data (in the factors or categorical predictors). ANOVA models can be analyzed through partitioning of sum of squares (an elegant traditional approach with the possibility of lots of additional explorations) or by reformulating them as regression models (computationally convenient approach). Both approaches are considered for the complete data, and then multiple imputation analysis under both approaches are presented. The ANOVA approach is a more general technique where the effects associated with factors could be treated as random or the factors could be nested or the effects of ordinal factors could be decomposed into various subcomponents. In all these situations, it will be useful to decompose the total variance into component variance and develop proper metrics to evaluate the magnitude of the components. Though, technically ANOVA can be formulated as a regression model and the underlying hypothesis expressed in terms of regression coefficients, there are advantages in dealing with the mean squares associated with various factors directly.

5.3.1 Complete Data Analysis

5.3.1.1 Partitioning of Sum of Squares

Consider a continuous outcome, Y, and two factors X_1 at J levels, X_2 at K levels. Suppose n_{jk}, $j = 1, 2, \ldots, J$; $k = 1, 2, \ldots, K$ is the sample size in the cell $X_1 = j, X_2 = k$. The standard two-way ANOVA model is

$$y_{ijk} = \mu + \alpha_j + \beta_k + (\alpha\beta)_{jk} + \epsilon_{ijk} \tag{5.1}$$

where $\sum_j \alpha_j = 0$, $\sum_k \beta_k = 0$, $\sum_j (\alpha\beta)_{jk} = \sum_k (\alpha\beta)_{jk} = 0$ and $\epsilon_{ijk} \sim$ *iid* $N(0, \sigma_\epsilon^2)$. The αs are the main Effects of X_1 or the deviations of the X_1 level means from the overall mean, the βs are the main effects of X_2 and $(\alpha\beta)$s are the interaction effects.

The analysis of variance splits the total sum of squares, $TSS = \sum_i \sum_j \sum_k (y_{ijk} - \bar{y}_{+++})^2$ where $\bar{y}_{+++}$ is the overall mean into four component sums of squares: Factor X_1, factor X_2, nonadditivity or interaction $(X_1 \times X_2)$ and the residual or unexplained sum of squares. For example, the factor X_1 sum of squares is

$$SS(X_1) = \sum_j n_{j+} (\bar{y}_{+j+} - \bar{y}_{+++})^2$$

where n_{+j} is the sample size for $X_1 = j$, and $\bar{y}_{+j+}$ is the mean of all the observations with $X_1 = j$. Similarly,

$$SS(X_2) = \sum_k n_{+k} (\bar{y}_{++k} - \bar{y}_{+++})^2,$$

and

$$SS(X_1 \times X_2) = \sum_j \sum_k n_{jk} (\bar{y}_{+jk} - \bar{y}_{+j+} - \bar{y}_{++k} + \bar{y}_{+++})^2$$

are the factor X_2 and the interaction $X_1 \times X_2$ sums of squares. The residual sum of squares is $RSS = TSS - SS(X_1) - SS(X_2) - SS(X_1 \times X_2)$. The degrees of freedom for the five sums of squares are $n-1$ for TSS, $J-1$ for $SS(X_1)$, $K-1$ for $SS(X_2)$, $(J-1)(K-1)$ for $SS(X_1 \times X_2)$, $\nu = (n-1) - (J-1) - (K-1) - (J-1)(K-1)$ for the residual sum of squares.

To assess the significance of the interaction effect, the ratio,

$$F_{12} = \frac{SS(X_1 \times X_2)/[(J-1)(K-1)]}{RSS/\nu} = \frac{mean\ square(X_1 \times X_2)}{residual\ mean\ square}$$

can be used which has an F distribution with $(J-1)(K-1)$ as the numerator degrees of freedom and ν as the denominator degrees of freedom when $(\alpha\beta)_{jk} = 0$ for all j, k (that is, no interaction between X_1 and X_2).

If the interaction effect is not significant then it makes sense to reduce the data further to the individual dimensions, X_1 and X_2. On the other hand, if the interaction effect is significant, then both dimensions have to be used jointly to summarize the data. That is, the effect of X_1 depends on X_2 (or the effect of X_2 depends on X_1). Further analysis of the means $\{\bar{y}_{+jk}, j = 1, 2 \ldots, J; k = 1, 2, \ldots, K\}$ may be needed to fully explore the joint impact these two factors on the outcome. For example, specific contrasts of substantive interest, orthogonal contrasts, response surface for nominal factors such linear, quadratic contrasts, etc. are some of the tools used in further investigation.

If the interaction effect is not significant then, strictly speaking, the model is refit without the interaction term or equivalently, combine the interaction and residual sum of squares as the new residual sum of squares. Defining $RSS^* = TSS - SS(X_1) - SS(X_2)$ which now has $\nu^* = n-1-(J-1)-(K-1)$ degrees of freedom. The F statistic used to assess the significance of the effect of X_1 is

$$F_1^* = \frac{SS(X_1)/(J-1)}{RSS^*/\nu^*}.$$

Many use

$$F_1 = \frac{SS(X_1)/(J-1)}{RSS/\nu}$$

which is more convenient (as the reduced model does not have to be fit). This is reasonable if the interaction sum of squares is rather small and X_1 and X_2 are orthogonal (like in a designed experiment). Similarly, F_2^* is constructed for X_2. If either factor is significant then further explorations using orthogonal contrasts or response surface for nominal factors will be undertaken.

5.3.1.2   Regression Formulation

The two-way ANOVA model can be formulated as a regression model with dummy variables. Let $A_j = 1$, if $X_1 = j$ and 0 otherwise, let J be dummy variables for the factor X_1. Similarly, $B_k = 1$, if $X_2 = k$ and 0 otherwise, K be dummy variables for factor X_2. Since $\sum_j A_j = 1$ (and $\sum_k B_k = 1$), only $J - 1$ and $K - 1$ dummy variables are needed when the intercept is included in the regression model. The following regression model is equivalent to the two-way ANOVA model:

$$y_{ijk} = \beta_o + \sum_{j=2}^{J} \beta_{1j} A_{ij} + \sum_{k=2}^{K} \beta_{2k} B_{ik} + \sum_{j=2}^{J} \sum_{k=2}^{K} \beta_{3,jk} A_{ij} B_{ik} + \epsilon_{ijk}.$$

In the above regression model, $X_1 = 1$ and $X_2 = 1$ are omitted and treated as reference categories. Each regression coefficient represents a contrast or comparison.

A test for the significance of the interaction effect is equivalent to assessing the magnitude of β_3 or testing for $\beta_3 = 0$ where $\beta_3 = (\beta_{3,22}, \beta_{3,23} \cdots, \beta_{3,32} \cdots \beta_{3,JK})^t$ is a $(J-1)(K-1)$-dimensional vector.

A partial F-test that compares the full model with the reduced model (omitting the interaction term) is equivalent to F_{12}. Another equivalent expression is $\widehat{D} = (\widehat{\beta}_3 - \beta_3^0)^t \widehat{V}^{-1} (\widehat{\beta}_3 - \beta_3^0)$ where β_3^0 (=0, in this case) is the null value, and $\widehat{V}$ is the covariance matrix of $\widehat{\beta}_3$. Advantage of this formulation is that β_3^0 can be any value dictated by the scientific question of interest. If the interaction effect is not significant, then the model can be reduced by dropping the product AB terms and then assess the significance of β_2 and β_1.

5.3.2   ANOVA with Missing Values

The regression analogy is useful to formulate the analysis plan with missing values. Consider three scenarios, the missing values in Y alone, in X alone and in both Y and X. If the missing values are in Y and the mechanism is MAR then the complete data method in the previous section can be applied. However, if an additional covariate Z is available then it may be advantageous to multiply impute the missing values.

If the missing values are in X (and possibly Y), clearly the missing values have to be imputed for both X and Y, and then the imputed values of Y may be discarded. It is important that the missing values are imputed without assuming any of the null hypothesis to be true. Otherwise the completed data will be biased towards the null. Thus, for the two-way analysis of variance model, imputation of Y needs to include X_1, X_2 and interaction $X_1 \times X_2$. Similarly, imputations of X_1 need to involve X_2, Y and $X_2 \times Y$ and imputations of X_2 need to involve X_1, Y and $X_1 \times Y$.

5.3.2.1 Combining Sums of Squares

At the basic level, all the tests in the analysis of variance involves two independent or approximately independent mean squares (sum of squares divided by its degrees of freedom). These mean squares are chosen such that their expected values are the same under a specific null hypothesis. For example, under the hypothesis of no interaction between X_1 and X_2, both numerator and denominator in F_{12} have the same expectation (σ_ϵ^2). Also, under the normal distribution of the residuals, a mean square MS on ν degrees of freedom with expectation σ^2 has the following property:

$$\frac{\nu \times MS}{\sigma^2} \sim \chi_\nu^2.$$

This is valid from both frequentist and Bayesian perspectives. From the frequentist perspective, the above equation is the sampling distribution of MS given σ^2 and from a Bayesian view, the equation is the posterior distribution of σ^2 given MS. Generally, such combination of a statistic and the parameter is call a "pivot." Another popular pivot is $V^{-1/2}(\beta - \widehat{\beta}) \sim N(0, I)$ where V is the covariance matrix, $\widehat{\beta}$ is an estimate and I is the identity matrix.

Let MS_l and ν_l be the mean square and degrees of freedom from the completed data $l = 1, 2, \ldots, M$. Noting that $E(\chi_\nu^2) = \nu$ and $Var(\chi_\nu^2) = 2\nu$, the completed data posterior mean of σ^{-2} is $e_l = 1/MS_l$ and the completed-data posterior variance of σ^{-2} is $U_l = 2/(\nu_l MS_l^2)$. The multiple imputation mean and variance estimates are

$$\widehat{\sigma}_{MI}^{-2} = \bar{e}_{MI} = \sum_l (1/MS_l)/M$$

and

$$T_M = 2 \sum_l (1/(\nu_l MS_l^2))/M + (1 + 1/M) \sum_l (1/MS_l - \widehat{\sigma}_{MI}^{-2})^2/(M-1),$$

respectively.

Using the approximation in equation (4.10), obtain

$$\sigma^{-2} \sim a\chi_b^2$$

or equivalently,

$$a^{-1}\sigma^{-2} \sim \chi_b^2$$

where $a = T_M/(2\bar{e}_{MI})$ and $b = 2\bar{e}_{MI}^2/T_M$.

In the ANOVA context, add D and N as the suffix for the numerator and denominator mean squares to obtain, $a_N^{-1}\sigma_N^{-2} \sim \chi_{b_N}^2$ and $a_D^{-1}\sigma_D^{-2} \sim \chi_{b_D}^2$, which leads to

$$F = \frac{\sigma_N^{-2}\, a_N^{-1} b_N^{-1}}{\sigma_D^{-2}\, a_D^{-1} b_D^{-1}},$$

having an approximate F-distribution with (b_N, b_D) degrees of freedom. Under the null hypothesis for which $\sigma_N = \sigma_D$, appropriate test statistic is

$$F = \frac{a_D b_D}{a_N b_N} = \frac{\bar{e}_{D,MI}}{\bar{e}_{N,MI}} = HM_N/HM_D$$

where HM is the harmonic mean of the completed data mean squares.

5.3.2.2 Regression Formulation with Missing Values

Section 4.6 covered multiparameter inference, which can be applied in the context of regression formulation. This involves computing appropriate covariance matrices, Eigenvalues, etc., a computationally complex procedure. Fortunately, the methodology described in Section 4.6 has been implemented in software packages such as PROC MIANALYZE.

5.3.3 Example

To illustrate both methods for performing ANOVA, consider data from wave 1 of the National Longitudinal Study of Adolescent Health (Add Health). The outcome variable Y is the vocabulary score and the two factors are household income X_1 (four categories based on the quartiles of the logarithm of income from complete cases) and the degree of positive attitudes towards sexual behavior X_2 (not positive, neither positive or negative, positive and highly positive based on an index constructed from five Likert type items). The total sample size for the subpopulation (excludes married subjects and also those less than 15 years of age) is $n = 4,154$. Missing values in X_1 is 1,034 (24.9%), 159 (3.8%) in X_2 and 182 (4.4 %) in Y. If one were to perform a complete case analysis by removing all subjects with missing values then the reduced sample size would be $m = 2,880$, or 69.4% response rate.

The missing values were imputed using the sequential regression approach including all the interactions as stated previously with Y imputed using a linear regression model, X_1 and X_2 using the multinomial logit model. A total of $M = 50$ imputations were used given a large amount of missing values.

Two-way analysis of variance with main effects for X_1 and X_2 and the interaction effects was performed on each completed data set and the necessary statistics (mean squares, degrees of freedom) were extracted. First, consider the test for interaction. The harmonic means of the numerator and denominator mean squares were 268.69 and 201.0, respectively, resulting in the F statistics $F = 1.34$. The total variance for the numerator and denominator were 7.2671×10^{-6} and 1.3369×10^{-8}, respectively, resulting in the degrees of freedom,

$$b_N = 2 \times 0.0037218^2 / 7.2671 \times 10^{-6} = 3.81$$

and

$$b_D = 2 \times 0.0049752^2 / 1.3369 \times 10^{-8} \approx 3703.$$

The p-value can be calculated as $Pr(F_{3.81,3703} \geq 1.34) = 0.254$.

Using the regression framework a model with three dummy variables for X_1, three for X_2 and their products (nine dummy variables) can be fit on each completed-data set, extract nine regression coefficients for the interactions and 9×9 matrices, $\bar{U}_M$ and B_M. The F-statistics in Section 4.6 and the degrees of freedom can be computed to test whether all nine regression coefficients for the interaction terms are equal to 0. This procedure is implemented in PROC MIANALYZE in SAS which was used in this example. The value of the test statistic was 0.62 and the numerator and denominator degrees of freedom were 9 and 3001.4, respectively, resulting in the p-value of 0.7830.

Given that the interaction between X_1 and X_2 is not significant, the reduced ANOVA and regression models were fit to assess the significance of the effects of X_1 and X_2. Table 5.4 summarizes the result from complete-case and the two multiple imputation analysis (ANOVA approximation and multiparameter test using the regression framework).

Both multiple imputation methods lead to essentially the same conclusion. The ANOVA method indicates a bit stronger interaction effect than

Table 5.3: Results from complete-case and two multiple imputation analysis of variance methods

Effect	F	DF	p-value
	Complete-Case		
$X_1 \times X_2$	0.72	(9,2866)	0.6901
X_1		(3,2875)	
X_2		(3,2875)	
	Regression Method		
$X_1 \times X_2$	0.62	(9,3001.4)	0.783
X_1	128.85	(3,1617.7)	< 0.0001
X_2	3.49	(3, 3210)	0.0151
	ANOVA Method		
$X_1 \times X_2$	1.34	(3.81,3703)	0.254
X_1	182.77	(3,3723)	< 0.0001
X_2	4.32	(2.43,3723)	0.009

the regression method which is close to the complete-case analysis. For many problems, it may be convenient to perform ANOVA method (such as nested designs, random effect models, etc.) where many software programs store the sums of squares and the degrees of freedom which are then amenable to simple calculations of means and variances.

5.3.4   Extensions

The ANOVA method described (see Section 5.3.4) for combining the mean squares can be used for conducting the partial F-test for a block of covariates in a multiple linear regression model. Suppose that R_{Fl} and R_{Rl} are the residual sums of squares from the full and reduced model from the completed data set $l = 1, 2, \ldots, M$. Define $MS_{Nl} = (R_{Rl} - R_{Fl})/(p_F - p_R)$ and $\nu_{Nl} = p_F - p_R$ where p_F (p_R) is the number of regression coefficients in the full (reduced) model. Define $MS_{Dl} = R_{Fl}/(n - p_F)$ and $\nu_{Dl} = n - p_F$ where n is the completed sample size.

The same approach can be used to perform the overall F-test in a regression analysis. Suppose that the regression model is $y = \beta_o + \sum_j \beta_j x_j + \epsilon$. Let M_l and R_l be the model and residual sum of squares from the completed data set $l = 1, 2, \ldots, M$. Define $\nu_{Nl} = p$, $\nu_{Dl} = n - p - 1$, $MS_{Nl} = M_l/p$ and $MS_{Dl} = R_l/(n-p-1)$. The ratio $F_{MI} = HM_N/HM_D$, of the harmonic means of the numerator and denominator mean squares is a test for the hypothesis $H_o : \beta_1 = \beta_2 = \ldots, \beta_p = 0$.

An interesting by-product is the multiple imputation combined R-square for the regression model. Using the well known relationship between R^2 and F in the complete data least squares analysis, define the multiple imputation R-square as

$$R^2_{MI} = b_N F_{MI}/(b_N F_{MI} + b_D)$$

and the multiple imputation adjusted R-square as $1 - (1 - R^2_{MI})(n - 1)/(n - p - 1)$.

5.4   Survival Analysis Example

This section presents a multiple imputation analysis of data from a clinical trial to evaluate the treatment for the primary biliary cirrhosis (PBC) of the liver. PBC is a rare but fatal chronic liver disease of unknown cause, with a prevalence of about 50-cases-per-million population. The study and the data set is described in the book, Fleming and Harrington (2005).

The study involves 424 PBC patients who were recruited for the randomized study. Only 318 agreed to be randomized to placebo or to the drug D-penicillamine (treatment) and the remaining 106, however, agreed to be followed to obtain the primary outcome: the survival time and subjects were censored at the time of liver transplantation, lost-to-follow up or the end of the study. Numerous covariates were measured at the baseline but not all were available on every subject. Most missing data were on subjects who did not agree to be randomized. Table 5.4 provides a description of the variables and the number of missing values.

The mean survival time estimated for the three groups (treated, placebo and observational) are 7.8, 7.5 and 6.9 years, respectively. The goal is adjust for covariates and compare the three groups. It is important to include the survival time as one of the predictors when imputing the covariates (as discussed previously in the CEX example). The censored cases are also missing outcome values (with censoring time providing the lower bound for their survival time). There are two possible ways in which the outcome can be incorporated in the prediction of the missing covariates: (1) Use censored/failure time and

Table 5.4: Description of the variables and the number of missing values. Censored observations are treated as missing values

Variable and Description	Number of Missing Values
Y: Survival time (log scale)	257
C: Censoring indicator (1=Censored, 0=Failure)	0
T: Censoring time	0
X_1: Treatment (1=D-Penicillamine 2=Placebo 3=Observational)	0
X_2: Age in years	0
X_3: Gender; 0=Male 1=Female	0
X_4: Presence of ascites (0=No 1=Yes)	106
X_5: Presence of hepatomegaly (0=No 1=Yes)	106
X_6: Presence of spiders (0=No 1=Yes)	106
X_7: Presence of edema (0=No 0.5=Edema-no treatment 1=Edema treatment)	0
X_8: Serum bilirubin (mg/dl) (log scale)	0
X_9: Serun cholesterol (mg/dl) (log scale)	134
X_{10}: Albumin (gm/dl)	0
X_{11}: Urine copper (mg/day) (log scale)	108
X_{12}: Alkaline phosphate U/liter (log Scale)	106
X_{13}: Serum glutamic-oxaloacetic transaminase (log scale)	106
X_{14}: Triglycerides mg/dl (log scale)	136
X_{15}: platelet count	11
X_{16}: Prothrombin time (in seconds)	2
X_{17}: Histologic stage (graded 1 to 4)	6

the censoring indicator variables as predictors in the model or (2) Jointly impute the missing values in the covariates and the missing failure times for the censored cases (treating the censoring time as the lower bound). The latter approach is more in tune with the Bayesian view of jointly imputing all the missing set of values conditional on the observed set of values. After the imputation, just like in the linear regression analysis, discard all the imputed values for the outcome variable (in the absence of any other good predictors of the outcome variable which are not part of the substantive model).

For this particular example, a normal regression model for the log survival time was used in the imputation, the log of the censoring time applied as the lower bound. The upper bound also applied (25 years) to avoid imputing large survival times (beyond reasonable age) for the censored subjects. The sequential regression approach was used to multiply impute the missing values for the covariates and the missing survival times ($M = 25$).

Model diagnostics showed that all the models fit the data well. For example, a scatter plot between any two variables, say (X_j, X_k), identified by a grouping defined as both observed values, X_j observed ("O") and X_k imputed ("I"), X_j imputed and X_k observed and both imputed is useful to assess whether the imputations appear to be plausible with respect to both dimensions. Several such plots were created to make sure that the imputations are plausible. Figure 5.2 provides two example scatter plots comparing observed and imputed values.

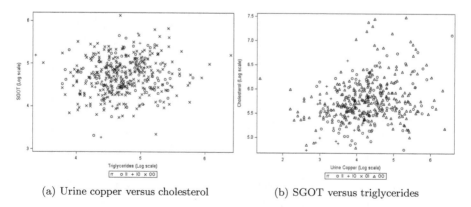

(a) Urine copper versus cholesterol (b) SGOT versus triglycerides

Figure 5.3: Diagnostic scatter plots to check the distribution of the observed and imputed values in the PBC analysis

After imputation, a Cox model for the hazard function was fit. The Cox model is defined by the hazard function,

$$\lambda(y) = \lambda_o(y)\exp(X^t\beta)$$

where $\lambda(y) = f(y)/(1 - F(y))$, $f(y)$ is the density function of the survival time y, $F(y) = \int_o^y f(u)du$ is the corresponding cumulative distribution function, $\lambda_o(y)$ is the unspecified (or arbitrary) baseline (or the control or reference population) hazard function which is modified by the covariates through the multiplicative factor, a function of the covariates.

The covariates were dummy variables for the group (treatment and observational and treating placebo as the reference category). All 16 covariates X_2 to X_{17} were included as predictors. Table 5.5 provides the estimated hazard ratios, their 95% confidence intervals and the fraction of missing information. The starred FMI correspond to covariates that are completely missing for the observational group. Note that the fully adjusted model cannot be fit on the complete-cases due to the starred missing values (which may be construed as 100% missing information).

Table 5.5: Results from the multiple imputation analysis of PBC data

Variable	Hazard Ratio	Lower 95% Limit	Upper 95% Limit	FMI
Treatment	0.954	0.646	1.407	5.2
Observational	1.082	0.667	1.754	12.3
X_2	1.034	1.015	1.053	6.5
X_3	1.095	0.625	1.918	10.1
X_4	1.397	0.719	2.714	33.0*
X_5	1.044	0.611	1.784	41.6*
X_6	0.957	0.595	1.539	34.4*
X_7	2.295	1.213	4.339	13.5
X_8	1.907	1.402	2.593	18.8
X_9	1.053	0.565	1.962	39.9*
X_{10}	0.613	0.382	0.982	8.1
X_{11}	1.473	1.074	2.022	28.6*
X_{12}	0.902	0.678	1.201	31.1*
X_{13}	1.506	0.833	2.724	33.3*
X_{14}	0.896	0.551	1.457	30.3*
X_{15}	1.074	0.644	1.79	21.4
X_{16}	1.218	1.039	1.428	11.3
X_{17}	1.413	1.075	1.857	11

There is not much difference between the three groups with the observational group slightly worse off than the placebo group and the treated group being slightly better off than the placebo group. Among the covariates, significant high risk factors are presence of edema, serum bilirubin, urine copper, platelet and the histologic stage of the disease.

5.5 Bibliographic Note

For a practice oriented book on regression analysis, see Gelman and Hill (2006). There are other classic books providing a comprehensive review of regression analysis such as Draper and Smith (1998), Weisberg (2013), Neter, Kutner, Nachtsheim, and Wasserman (1996).

Some of the early work on missing values in regression analysis is described in Afifi and Elashoff (1967) and for the missing values in ANOVA, Allan and Wishart (1930). See also Dodge (1985). Rubin (1976b, 1976c) discusses missing values in the outcome and predictors. A comprehensive review of regression analysis with missing predictors is given in Little (1992). More recent references include Von Hippel (2007).

The consumer expenditure survey simulation example is described in Raghunathan and Paulin (1998). For more information about Tobit models, see Amemiya (1984) and for survival analysis, see Fleming and Harrington (2005) and Kalbfleisch and Prentice (2002).

5.6 Exercises

1. Consider data from the St. Louis risk research study. Create your own set of multiply imputed data sets with G, V and R variables. From the imputed data sets, create paired differences $Y_{1j} = V_{1j} - V_{2j}$ and $Y_{2j} = R_{1j} - R_{2j}$, $j = 1, 2, \ldots, 69$.

    (a) Perform a complete case analysis by regressing Y_1 (Y_2) on dummy variables G_2 and G_3.

(b) Perform a multiple imputation version of the two regression analyses in the previous problem.

(c) Use a random effects model to jointly analyze the bivariate outcome (Y_1, Y_2).

(d) Use the results described in Section 5.2 and the current analysis to write a report on your findings. Specifically address the merits and demerits of the current analysis with reference to the analysis described in Section 5.2.

2. Download data from the most recent National Health and Nutritional Examination Survey (NHANES) that has the following variables: education, body mass index (BMI), gender, age, race/ethnicity and fasting glucose and hemoglobin A1c. Create four categories of education: less than high school, high school, some college and completed college; three categories of obesity: BMI less than or equal to 25, BMI greater than 25 and less than 30 and BMI greater than or equal to 30; race/ethnicity into five categories: non-hispanic white, non-hispanic black, mexican, other hispanics and other races; and, finally, categorize age into three categories: 25-34, 35-44 and 45-54.

(a) Perform multiple imputation analysis of variance with Hemoglobin A1c as the outcome variable with education, age and gender as factors. Prepare the analysis of variance table and appropriate F-tests. Make sure to explore any interaction effects among the factors.

(b) Formulate the same analysis using a regression and carry out the tests for the same hypotheses corresponding to the ANOVA table entries.

(c) To the model in the previous problem add race/ethnicity as a factor and refine the model by dropping or adding interaction effects.

(d) Finally, now add obesity as a factor and refine the model.

(e) Based on these analyses, write a report on the role of all the five factors on hemoglobin A1c.

3. In the previous problem, some continuous variables were categorized. Perform a regression analysis with a mix of categorical and

continuous predictors to investigate the effect of education, BMI, age, gender and race/ethnicity on hemoglobin A1c.

4. Repeat the analysis in problems 2 and 3 but now using fasting glucose as the outcome variable without using hemoglobin A1c as a "study" variable.

5. Redo problem 4 but using hemoglobin A1c as an auxiliary variable. How will you assess the utility of using hemoglobin A1c when studying the impact of predictors on fasting glucose?

6. **Project.** Generate 500 data sets, each of size 200, from the following set of models $x \sim N(\mu, \theta^2)$, $y|x \sim N(\alpha_o + \alpha_1 x, \sigma^2)$ and $z|y, x \sim N(\beta_o + \beta_1 x + \beta_2 y, \tau^2)$. The goal is to obtain the point and interval estimate of α_1. Set some values of x to missing using the response propensity model $\text{logit} Pr(x \text{ is Missing}|z, y) = \gamma_o + \gamma_1 y + \gamma_2 z)$ and set some values of y to missing, only if x is not missing, using the response propensity model $\text{logit} Pr(y \text{ is Missing}|z, x) = \delta_o + \delta_1 x + \delta_2 z)$. Choose γs and δs to generate about 40% complete cases, 20% missing y and 20% missing x.

   (a) Analyze 500 data sets before setting x or y values to missing and store the point and interval estimates of α_1.

   (b) Perform complete-case analysis of 500 data sets after setting some x or y values to missing and store the point and interval estimates of α_1.

   (c) By varying the choices of γs and δs explore the bias in the complete-case analysis. Based on this exploration choose three sets of parameters (γs and δs) corresponding to your own subjective definition of "low," "medium" and "large" bias situations.

   (d) For each set of parameters perform two sets of multiple imputations (you may choose the number of imputations, M, in 10 to 20 range): (1) Focus on (y, x), ignoring z; and (2) Focus on (y, x, z) for the imputation.

   (e) Perform 2 sets of analysis for each set of imputations: (1) Ignore the imputed values of y in the regression analysis of y on x, and (2) Include the imputed values of y in the analysis.

(f) This simulation results in six point estimates for each of the 500 data sets: These are before Deletion, complete-case and 4 multiple imputation versions. Summarize the sampling distribution of these estimates as a function of factors: Bias situation (three levels), use/ignore auxiliary information in z (two levels) and use/ignore the imputed values in y (two levels).

(g) For the six sets of interval estimates compute the length and coverage rates (defined as the percentage of intervals containing the true value of α_1). Summarize the distribution of lengths and coverage rates as a function the same factors listed in (f).

(h) Based on this project write a succinct summary of recommendations on a regression analysis of data with missing values in the outcome and predictor variables.

6

Longitudinal Analysis with Missing Values

6.1 Introduction

A longitudinal study collects measurements from the same set of individuals repeatedly over time. This could be in a randomized clinical trial with intense follow up for a short duration or in a panel survey with long term follow up and a year or more gap between measurements. Analysis becomes complicated when not all subjects participate at every time point, either by design or by choice. Ignoring subjects who dropped out may bias some results. For example, if the goal is to estimate the mean difference at the end of the study and project it to a population, then subjects who are alive and lost to follow-up need to be included. There may be some analysis where the population of completers may be of interest. If the completers, however, are highly selected then the analysis restricted to them may not be meaningful. Clearly, people with missing covariate data cannot be ignored.

The longitudinal data with p waves or periods of data collection can be structured in a "wide format" with one row per person with the covariates X, and all the measures $Y_1, Y_2, \ldots, Y_p$ "strung out." Note that each Y_i may be a row vector of variables. In many longitudinal studies, the follow-up may not happen exactly at the same time for all the individuals. For example, a study protocol may expect visits to occur every month but due to scheduling delays or for some other reasons, the respondents may be late for some visits. The time between the visit for every individual is typically included in the data set. The time of measurement is missing for those who are lost to follow-up. In any case, the time between the follow-up visits has to be included as a covariate, if it is not constant across all the individuals.

An alternative, and perhaps the most useful, is the "long format," which has multiple rows per subject with the number of rows equal to the number of follow-ups. Most computer software for analyzing the longitudinal data require the data in the long format due to its flexibility in handling unbalanced (that is, missing outcome or dependent variable) data. In the multiple imputation analysis of longitudinal data, imputation process may need the data in the wide format, but the analysis of completed data may need the long format.

Consider an example with data on n subjects collected at four time points, Y_1, Y_2, Y_3 and Y_4 and one covariate X. The missing values can be in the outcomes or the covariate. A growth curve model is typically used for the analysis of longitudinal data to assess the effect of time and the differential effect of time across covariate groups. To be concrete, consider the following random effects growth curve model,

$$Y_{it} = \alpha_i + \beta_i t + \epsilon_{it}, \tag{6.1}$$

where $t = 1, 2, 3, 4$; $i = 1, 2, \ldots, n$,

$$\begin{bmatrix} \alpha_i \\ \beta_i \end{bmatrix} = \begin{bmatrix} \theta_0 + \theta_1 x \\ \theta_2 + \theta_3 x \end{bmatrix} + \begin{bmatrix} \eta_{0i} \\ \eta_{1i} \end{bmatrix},$$

$$\eta_i = \begin{bmatrix} \eta_{0i} \\ \eta_{1i} \end{bmatrix} \sim N \left(\begin{bmatrix} 0 \\ 0 \end{bmatrix}, \Omega \right),$$

$\epsilon_{it} \sim N(0, \sigma^2)$, Ω is a 2×2 matrix, and ϵs and ηs are all mutually independent.

When X has missing values, a model is needed to construct a conditional predictive distribution of the missing values. Assume that $X \sim N(\mu, \tau^2)$. Thus, the joint distribution of (Y_1, Y_2, Y_3, Y_4, X) can be reexpressed as

$$\begin{bmatrix} Y_{i1} \\ Y_{i2} \\ Y_{i3} \\ Y_{i4} \\ X \end{bmatrix} \sim N \left(\begin{bmatrix} \theta_0 + \theta_1 \mu + \theta_2 + \theta_3 \mu \\ \theta_0 + \theta_1 \mu + 2\theta_2 + 2\theta_3 \mu \\ \theta_0 + \theta_1 \mu + 3\theta_2 + 3\theta_3 \mu \\ \theta_0 + \theta_1 \mu + 4\theta_2 + 4\theta_3 \mu \\ \mu \end{bmatrix}, \Sigma \right)$$

where Σ is a known function of σ^2, τ^2 and Ω. The mean function involves five unknown parameters $(\mu, \theta_0, \theta_1, \theta_2, \theta_3)$ and five parameters in the covariance matrix, Σ. If the timing of measurements is not the same for all the individuals then t in equation (6.1) is also missing for those with missing Ys. The missing values in t may be set by randomly choosing from the observed set of values to reflect the distribution of observed time points among the imputed values. For now assume that the timing of measurements is the same for all subjects in the study.

A possible approach for handling missing data in longitudinal studies is to structure the data in the wide format (in this case, five variables on each subject) and then multiply impute the values using the standard methods described in Chapter 3. A question arises, "Are the imputations obtained using the suggested method valid if the growth curve model were to be used for the analysis?" To investigate further, suppose that the imputation model uses a multivariate normal model with mean as a five-dimensional vector and an arbitrary covariance matrix which involves 15 unknown parameters. Thus, the imputation model involves more parameters in the covariance matrix or, equivalently, less structure than in the analysis model.

In a general setup with p time points, the number of parameters in the growth curve model remains the same (five for the mean and five for the covariance matrix) but increases to $(p+1)(p+4)/2$ for the multivariate normal multiple imputation model ($(p + 1)$ for the mean vector and $(p + 1)(p + 2)/2$ for the covariance matrix). That is, the imputation model is more general (or assumes less structure) than the analysis model.

In the sequential regression approach, $p + 1$ regression models are used, each with a minimum of $p + 1$ predictors including the intercept (addition of nonlinear and interaction terms may increase the number of regression coefficients) and a residual variance. Thus, the total number of parameters used in the imputation model is at least $(p+1)(p+2)$ which is more than the number of parameters in the unstructured multivariate normal assumption described above. Thus, in terms of the number of parameters, the growth curve model is a "sub-model" of the multivariate normal imputation model, which in turn is a "sub-model" of the sequential regression approach, but all assume normality.

There can be two points of view based on the degree of confidence in growth curve model assumption. If the growth curve model is the best fitting model (or very reasonable model based on diagnostics) or it is treated as the true model under the frequentist framework for the repeated sampling calculations then both imputation models (multivariate normal or sequential regression) are introducing noise through estimating additional parameters, which should actually be set to zero or some other fixed value. This, generally, will lead to an increase in the sampling variance of the multiple imputation estimate, or for that matter, any statistic constructed from the multiply imputed data set.

On the other hand, if one does not want to treat the growth curve model as "the gold standard" but rather a convenient or simplified representation of the relationship, then the less restrictive imputation model may be preferable for constructing plausible data sets. Expanding beyond normality in the imputation process may be preferable as well. Of course, the sample size and the number of variables and the amount of missing values will be constraining factors while developing a good fitting imputation model. In general, one may prefer to impute the missing values under much less restrictive assumption than any of the analysis model to be used subsequently. This point of view is perhaps more important in a setting where the imputer and the analysts are different (for example, public use data files, multi-center trials, etc.) when compared to the scenario where the analyst is using multiple imputation as a tool to analyze the data under a set of model assumptions which is assumed to be true.

The imputation task is challenging when the number of variables becomes very large, even greater than the number of subjects in the study, in the horizontally concatenated data set. Modeling tasks can become difficult but the suggestions made earlier (such as using the principal components of the predictors) could be used to reduce the dimensionality of the predictors in the imputation models.

6.2   Imputation Model Assumption

The most important issue, however, is the assumption about the conditional distribution of the missing observations for subjects who have dropped out, given the observed data. To be concrete consider a randomized trial with two arms $T = 1$ (treatment) and $T = 0$ (placebo), with a protocol of p follow-ups, measuring an outcome variable Y at each visit. Let X be a vector of variables measured at baseline, before randomization. There are many ways to evaluate the treatment and, for concreteness, suppose that the parameter of interest is $\psi = E(Y_p|T = 1, X) - E(Y_p|T = 0, X)$, the mean difference between the two treatment groups at the last visit adjusted for the baseline covariates X.

When subjects, who drop out at visit k, never come back, the missing data pattern is monotone. In some studies, subjects may participate intermittently leading to an arbitrary pattern of missing data. For now, assume a monotone pattern of missing data. An important questions is "What assumptions should one make when estimating ψ about the people who are not observed at time point p"?

The first scenario may be called "completed as randomized." Consider two individuals, i and j, where i completed the study but j dropped out at time k. Partition $y = (y_1, y_2, \ldots, y_{k-1}, y_k, y_{k+1}, \ldots, y_p) = (Y_{(k-1)}, Y_{(k)})$ where $Y_{(k-1)}$ denotes observations up to and including time $k-1$, and $Y_{(k)}$ denotes observations from time k to p. Under this assumption,

$$Pr(Y_{j,(k)}|Y_{j,(k-1)} = z, X = x, T = t) \equiv Pr(Y_{i,(k)}|Y_{i,(k-1)} = z, X = x, T = t).$$

This is also the standard missing at random assumption (that is, missingness depends only on the observed data for each subject). Imputations carried out under this scenario, projects plausible data sets of completers as randomized.

The second scenario for imputations may be called as "completed as control." Here all the future missing values on subjects who dropped out at time k are imputed as though they are control subjects from time period k onwards, but conditioning on the observed values. The imputation model assumes that

$$Pr(Y_{j,(k)}|Y_{j,(k-1)} = z, X = x, T = t) \equiv Pr(Y_{i,(k)}|Y_{i,(k-1)} = z, X = x, T = 0).$$

The third scenario for imputation may be called "completed as stable." Here all the future missing values on subjects who dropped out at time k are imputed as their last known value plus residual where the residual is drawn from an estimated distribution based on the observed data. This approach may be viewed as a more principled version of an ad hoc approach, the last-observation-carried-forward (LOCF) method, where, for any given subject, the last known value is imputed for all the observations in the future. The LOCF approach fails to recognize variation around the actual measurements whereas the proposed method does. The imputation model is

$$Y_{jl} = y_{j,k-1} + \epsilon_{jl}$$

where, for $l = k, k+1, \ldots, p$, $\epsilon_{jl} \sim \widehat{F}_{obs}$, the predictive distribution of the residuals, conditional on the observed data on all the subjects.

One can imagine several other scenarios. The scenarios, "completed as randomized," "completed as control" and "completed as stable," are just a few examples of the projections of plausible completed data sets to provide a range of options to assess the effect of treatment. There are other possible scenarios such as "completed as decliners" or "completed as improvers."

The choice of scenarios depends upon the scientific question of interest. For example, the goal of the "completed as randomized" imputation strategy is to obtain the effect of treatment under an ideal situation; "completed as control" imputation strategy goal is to truly assess the efficacy of the treatment in a population where some people will drop out and, consequently, will behave like controls (immediately after drop out). Finally, "completed as stable" imputation strategy assumes that accrual of benefit will cease after the drop out, except for some random measurement error. The other scenarios may provide information about the effect of treatment in the "field" conditions where some people will drop out and as a consequence outcomes may shrink towards the standard or worse or even better. It may be prudent to perform several such analyses to develop insight into potential and realized effect of intervention. Performing several plausible analyses is also useful to assess the robustness of the findings about the treatment effect with respect to missing data mechanisms.

6.2.1   Completed as Randomized

Under MAR assumption, with no missingness in the covariates and normally distributed outcome measures, the maximum likelihood (ML) approach is more common than multiple imputations (although, MI may have certain advantages as discussed in context of an example in Section 6.3). Many software packages have implemented ML (for example, PROC MIXED in SAS). When the covariates have missing values then the current software packages ignore subjects with missing values.

Let $\phi(y|\theta, \Sigma)$ be the multivariate normal density for the complete data $y = (y_1, y_2, \ldots, y_p)$ with a p dimensional mean vector θ which may be a function of the covariates X and a $p \times p$ covariance matrix, Σ, which may also be a function of covariates. Let y_{obs_i} be a vector of the observed values for subject i where obs_i denotes a subset of $\{1, 2, \ldots, p\}$ corresponding to the observed time points for the subject i.

The marginal density of y_{obs_i} is also multivariate normal with mean $\theta_i = \theta_{obs_i}$ (a subvector of θ) and $\Sigma_i = \Sigma_{obs_i, obs_i}$ (a submatrix of Σ). The likelihood from subject i, $L_i(\theta_i, \Sigma_i|y_{obs_i})$, is the multivariate normal density evaluated at the numerical values of the observed data on subject i (a function of θ_i and

Σ_i). Thus, the observed data likelihood across all n subjects in the study is

$$L_{obs}(\theta, \Sigma) \propto \prod_i^n L(\theta_i, \Sigma_i | y_{obs_i}).$$

The likelihood maximized with respect to θ and Σ results in the maximum likelihood estimates, $\widehat{\theta}$ and $\widehat{\Sigma}$. Maximization involves iterative numerical optimization routines such as the Newton-Raphson method or several other variants. Though the maximum likelihood approach described here assumed normality, technically any distribution could be used as long as the marginal density functions of y_{obs_i} can be calculated. But software for other distributions is not readily available.

The maximum likelihood estimate of any other function $\psi = \psi(\theta, \Sigma)$ is obtained as $\widehat{\psi} = \psi(\widehat{\theta}, \widehat{\Sigma})$. The standard errors of $\widehat{\psi}$ can be computed using the so-called delta method. Another option is to use the Bootstrap approach, as described below.

1. Sample n subjects with replacement from the original sample of size n.

2. Obtain an estimate of ψ based on the sample in step (1).

3. Repeat steps (1) and (2) a total of B times.

Let $\widehat{\psi}_b$ be the estimate from the bootstrap sample $b = 1, 2, \ldots, B$. The bootstrap estimate of the standard error of $\widehat{\psi}$ is given by

$$SE_B^2(\widehat{\psi}) = \sum_b (\widehat{\psi}_b - \bar{\psi}_B)^2 / [(B(B-1)],$$

where $\bar{\psi}_B = \sum_b \widehat{\psi}_b / B$ is the bootstrap estimate.

An alternative approach for obtaining the standard errors is based on the jackknife approach as described below.

1. Create n replicates each of size $(n-1)$ subjects by deleting one individual at a time.

2. Let $\widehat{\psi}_r$ be the estimate from replicate $r = 1, 2, \ldots, n$.

3. Define $\psi_r^* = n\widehat{\psi} - (n-1)\widehat{\psi}_r$, which are called pseudo values.

4. The jackknife bias corrected estimate is $\bar{\psi}_J^* = \sum_r \psi_r^* / n$.

5. The standard error is given by the formula,

$$SE_j^2(\bar{\psi}^*) = \sum_r (\psi_r^* - \bar{\psi}_j^*)^2/[n(n-1)]$$

$$= (n-1)\sum_r (\widehat{\psi}_r - \widehat{\psi})^2/n.$$

Both these approaches are easy to implement in many software packages using the built-in functions.

When some covariates are missing, the likelihood approach becomes difficult to implement. A two stage approach could be used where the missing values in the outcome and the covariates are multiply imputed at the first stage but the imputed outcomes are ignored at the end of the imputation process. At the second stage, ML approach is used on each covariate-completed data sets, the results are then combined using the standard combining rules.

Though maximum likelihood is the standard approach, multiple imputation can also be performed, then analyze each completed data set and combine point estimates, covariance matrices, test statistics, etc. to form a single inference. This approach may be appealing where multiple outcomes are collected but not all are measured at every time point. For example, suppose that blood pressure, cholesterol, plasma glucose are the outcome variables. But not all are measured at every time point on every subject. It may be beneficial to exploit the correlation between these variables to impute the missing values jointly even though, say, blood pressure is the analysis variable of interest.

6.2.2 Completed as Control

Completed as control strategy borrows strength from other controls to impute the missing outcomes for the subjects in the treatment group immediately following the drop out. An easy way to implement the multiple imputation approach is as follows. Define time varying variables, $T_1, T_2, \ldots, T_p$, to indicate the treatment group assignment for time points. Here $T_1 = T$, the original treatment assignment. In wave two, all subjects who dropped out of the treatment group will be switched over to the control.

Specifically, $T_{i2} = T_{i1}$, if the subject i continues to be in the study. If the subject i is a drop out, then $T_{il} = 0$ for $l = 2, 3, \ldots, p$. In general, all subjects that dropped out at time k (that is $y_{i,k-1}$ is observed) will be switched over as control at time $k, k+1, \ldots, p$ and keep the original assignment if they continue to be in the study. The data set has now p more variables which are used as covariates in the imputation model.

Further refinement is to stratify the data based on the treatment assignment. The dropouts from the treatment group are transferred to the control group at the time of drop out. The imputations may then be carried out separately for the two treatment groups. Separate imputation strategy preserves the independence of observations on subjects in the two different groups. An advantage of the separate imputation strategy is that the treatment effect or the effect of any other covariate is not "smoothed" across the two groups but may be difficult to implement for small-scale studies. A compromise approach involves prediction or imputation modeling with interaction terms between the treatment assignment and other predictors in order to preserve the heterogeneity of effects.

6.2.3 Completed as Stable

Completed as stable scenario requires estimating the residual variance, $\widehat{\sigma}_l^2$, which can be obtained from the "completed as randomized" analysis for time period l and then define imputations as $y_{il}^* = y_{i,k-1} + z_{il}\widehat{\sigma}_l$ for $l = k, k+1, \ldots, p$ where z_{il} is a random normal deviate.

A simpler method is to use the analysis of variance model (within each treatment group),

$$y_{il} = \mu + \alpha_l + \gamma_i + \epsilon_{il}$$

where y_{il} is the observation on subject i, at time l, α_l is the time effect , γ_i is the subject effect and ϵ_{il} is the residual assumed to be normally distributed with mean 0 and variance σ_ϵ^2. One can obtain estimates, $\widehat{\mu}$, $\widehat{\alpha}_l$, $\widehat{\gamma}_i$ and $\widehat{\sigma}_\epsilon^2$. The imputed values are then defined as

$$y_{il}^* = y_{i,k-1} + z\widehat{\sigma}_\epsilon$$

for $l = k, k+1, \ldots, p$. To account for estimation of the residual variance, use the perturbed value,

$$\sigma_\epsilon^* = \sqrt{\nu\widehat{\sigma}_\epsilon^2 / \chi_\nu^2},$$

instead of $\widehat{\sigma}_\epsilon$, where ν is the degrees for the error sum of squares, and χ_ν^2 is a chi-square random variable with ν degrees of freedom.

6.3   Example

A randomized clinical trial was conducted to evaluate buprenorphine-naloxone (Bup-nx) versus clonidine for opioid detoxification. A total of 113 in-patients passing a screening test and satisfying inclusion criteria were randomized with 77 patients receiving Bup-nx and 36 receiving clonidine. The study involved measuring patients for 14 days with a base-line measure coded as day 0 (before the treatment began). A variety of measures were collected to assess the treatment effect and one of them is a self-report measure, visual analog scale (VAS) based on the question, how much do you currently crave for opiates? with a continuous response options ranging from 0 (no cravings) to 100 (most extreme cravings possible).

Not every patient responded to this question on all days. Table 6.1 provides the number, mean and the standard deviation of measurements for each day by treatment group. Very few patients completed the study and the drop out is more severe in the clonidine group. One of the goals is to compare the responses on day 14 between the two groups, adjusted for the covariates: age, gender, race/ethnicity.

Figure 6.1(a,b) provides box plots of the observed responses over the 14 days for the two treatment groups. There is a considerable amount of variation across individuals, but the medians are suggestive of a quadratic relationship in both groups. The means for the clonidine group show a pattern similar to Bup-Nx group, except around Day 7, where there is an increase and then a decline around day 10.

6.3.1   Completed as Randomized: Maximum Likelihood Analysis

For the "completed-as-randomized" analysis, based on Figure 6.1 (a,b) and some preliminary analysis, the following growth curve model may be used,

$$y_{id} = \beta_{0i} + \beta_{1i}d + \beta_{2i}d^2 + \epsilon_{id}$$

where

$$\beta_{0i} = \theta_{00} + \theta_{01}T + \theta_{02}\text{Age} + \theta_{03}\text{Female} + \theta_{04}\text{White} + \eta_{0i},$$

$$\beta_{1i} = \theta_{10} + \theta_{11}T + \eta_{1i},$$

and

$$\beta_{2i} = \theta_{20} + \theta_{21}T + \eta_{2i}.$$

Table 6.1: Mean and standard deviation of visual analog scale score by treatment group and days of treatment

Day	Bup-Nx (Treat=1)			Clonidine (Treat=0)		
	n	Mean	SD	n	Mean	SD
0	75	61.5	30.3	35	65.4	28.8
1	64	64.5	28.2	32	66.8	26.3
2	74	46.0	28.6	29	58.3	34.2
3	72	35.3	30.5	27	46.7	32.2
4	71	26.3	26.1	20	33.9	27.6
5	71	24.3	27.2	17	26.4	30.0
6	67	18.6	21.2	18	18.7	20.2
7	70	18.3	21.4	14	12.6	17.2
8	69	16.1	19.0	12	19.5	28.1
9	65	14.0	17.2	12	21.1	29.0
10	63	13.2	17.3	10	20.9	31.5
11	63	9.7	12.6	9	16.3	31.5
12	60	12.8	18.8	8	5.0	6.4
13	62	12.4	18.9	8	4.5	5.3
14	57	11.8	19.3	8	9.3	10.1

The within-subject error terms $\epsilon_{id} \sim N(0, \sigma_\epsilon^2)$, the "between-subject" error terms $(\eta_{0i}, \eta_{1i}, \eta_{2i})$ has a trivariate normal distribution with mean $(0, 0, 0)$ and

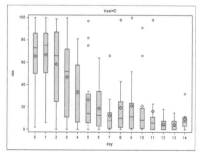 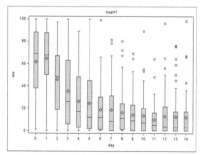

(a) Box plot of daily visual analog scale measures from the respondents in the clonidine group (Treat=0)

(b) Box plot of the daily analog scale measures from the respondents in the Bup-Nx group (Treat=1)

Figure 6.1: Longitudinal study comparing Bup-Nx and clonidine for detoxfication

Table 6.2: Maximum likelihood estimate of the parameters

Parameter	Estimate	SE
Intercept(θ_{00})	66.82	8.11
Bup $-$ Nx(θ_{01})	-7.24	5.46
Age(θ_{02})	0.022	0.165
Female(θ_{03})	3.21	3.34
White(θ_{04})	3.43	3.32
Day(θ_{10})	-8.21	1.66
Bup $-$ Nx $\times$ Day(θ_{11})	-1.36	1.92
Day^2(θ_{20})	0.299	0.121
Bup $-$ Nx $\times$ Day^2(θ_{21})	0.146	0.137
ψ	2.23	6.98

a 3×3 covariance matrix Ω. All the error terms are independently distributed. The treatment dummy variable, T, is set to 1 for Bup-Nx and 0 for clonidine.

Under this model, the parameter of interest is the covariate adjusted difference in the means between the two groups on day 14, $\psi = \theta_{01} + 14\theta_{11} + 196\theta_{21}$. The maximum likelihood estimate and its standard error can be obtained using a software package such as PROC MIXED in SAS. The "ESTIMATE" feature in PROC MIXED makes it easy to infer about ψ.

Table 6.2 provides the maximum likelihood estimates of the regression coefficients and ψ and their standard errors. At baseline, the age/race/gender adjusted difference between the Bup-Nx and clonidine group is about -7.24 with the standard error 5.46 and by the end of the study the estimated difference is 2.23 with the standard error 6.98. Neither difference is statistically significant.

The maximum difference over the 14 day period may be obtained by analyzing the temporal difference between the two groups $\psi(t) = \theta_{01} + \theta_{11}t + \theta_{21}t^2$ which reaches maximum at t^*, the solution to the equation $d\psi(t)/dt = \theta_{11} + 2t\theta_{21} = 0$ or $t^* = -\theta_{11}/(2\theta_{21})$. The estimate of t^* is $1.3621/(2 \times 0.1456) = 4.7$ days and the corresponding maximum difference is -10.42. That is, patients in the Bup-Nx group score about 10 point less on VAS score than those in the clonidine group. This measures seems to be subject considerable within-subject variation ($\hat{\sigma}_\epsilon^2 = 249.25$) as well as the between-subject variation as indicated by the estimated Ω matrix,

$$\begin{bmatrix} 580.79 & -97.45 & 4.83 \\ -97.45 & 52.77 & -3.46 \\ 4.83 & -3.46 & 0.24 \end{bmatrix}.$$

6.3.2 Multiple Imputation: Completed as Randomized

In a mutlivariate normal setup, the data set has 113 rows and 19 variables (3 covariates: age, gender and race; a treatment indicator and 15 possible response variables). A sequential regression multivariate imputation approach was used to impute the missing values. The number of observed values in this data set is $m = 1,262$, out of possible $n = 113 \times 15 = 1,695$ (that is, the missing data percentage is 25.6%). Figure 6.2(a,b) provides box plots of the multiply imputed outcome variables based on $M = 25$ imputations. The box plots show similar patterns as with the observed data but the clonidine group shows a cyclical pattern when compared to the Bup-Nx group.

After imputation, the completed data sets were analyzed using an unstructured model with the mean function,

$$y_{it} = \alpha_o + \alpha_1 T_i + \alpha_2 \text{Age}_i + \alpha_3 \text{Female}_i + \alpha_4 \text{White}_i +$$

$$\sum_{j=5}^{19} \alpha_j D_{it} + \sum_{j=20}^{34} \alpha_j T_i D_{it} + \epsilon_{it}, \qquad (6.2)$$

where D_{it} is a dummy or indicator variable for day $t = 1, 2, \ldots, 14$ and $\epsilon_i = (\epsilon_{i0}, \epsilon_{i1} \ldots, \epsilon_{i,14})^t$ having a multivariate normal distribution with mean 0 and an arbitrary 15×15 variance-covariance matrix Σ. This model may be a bit of a stretch for this data given the limited sample size.

Table 6.3 provides the estimated mean difference between the two treatment groups for each day, its standard error and the fraction of missing information. Generally, subjects in the Bup-Nx group are reporting less cravings

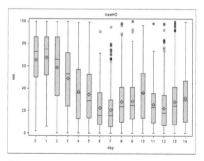 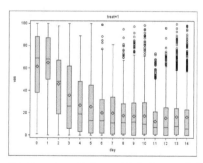

(a) Box plot of the multiply imputed daily visual analog scale measures for the subjects in the clonidine group (Treat=0)

(b) Box plot of the multiply imputed daily analog scale measures for the subjects in the Bup-Nx group (Treat=1)

Figure 6.2: Longitudinal study comparing Bup-Nx and clonidine for detoxfication

Table 6.3: Covariate adjusted multiple imputation mean difference between the Bup-Nx and clonidine groups for each day of the study, its standard error and the fraction of missing information

Day	Estimate	SE	FMI (%)
0	-4.20	6.08	1.2
1	1.13	5.07	10.8
2	-8.77	7.88	7.6
3	-9.31	8.23	10.8
4	-6.23	8.00	11.0
5	-5.11	8.17	14.6
6	1.59	7.38	8.6
7	2.97	7.59	14.9
8	-6.30	8.11	24.8
9	-7.35	7.15	13.6
10	-14.97	8.20	30.5
11	-8.86	7.44	19.2
12	-2.92	8.26	36.0
13	-7.48	8.14	29.5
14	-11.26	10.20	54.7

for opiates than those in the clonidine group but none of the differences are statistically significant.

Figure 6.3 provides a plot of mean differences between treatment groups over the 14-day period. This plot reveals a cyclical pattern of the differences

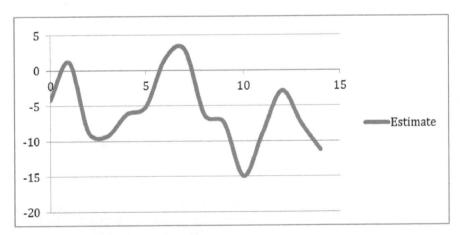

Figure 6.3: Estimated daily mean difference between Bup-Nx and clonidine groups

with a downward trend. A quadratic function for the mean similar to the growth curve model but with an unstructured 15×15 covariance matrix was also fit to estimate ψ, the expected mean difference between the two groups on day 14. The multiple imputation estimate was $\widehat{\psi}_{MI} = -4.06$ with the standard error 7.38.

6.3.3 Multiple Imputation: Completed as Control

For this analysis, the missing values are imputed by drawing values from the predictive distribution under the standard or control treatment, clonidine. To do so, define time-varying treatment variables which assign all missing values to the clonidine group. Specifically, $T_{it} = T_i$ is y_{it} is observed and $T_{it} = 0$ is y_{it} is missing. Multiple imputation using a sequential regression approach was carried out on the augmented data set.

With $M = 25$ imputations, the same analysis using the model in equation (6.1) was carried out. The covariate adjusted difference in the day 14 means between those assigned to Bup-Nx and clonidine groups is -10.22 (with standard error 10.19), only slightly smaller than -11.26 in Table 6.4.

6.4 Practical Issues

In panel surveys, often the data from each wave are released for the analysis soon after completing the data collection. Suppose that the first wave data collection has ended, the missing values are multiply imputed and released to the public. When the second wave data is collected, then the missing values in both the waves have to be jointly imputed in order to maintain the bidirectional relationship between variables across the two waves. This implies that wave one imputations will change. As the time progresses, all previous and current waves have to be reimputed and imputed, respectively, jointly in order to maintain the "panel property" among the imputed values.

A key challenge, therefore, in the imputation of variables in a longitudinal study is the exploding number of variables with the number of follow-ups. Mass imputation of variables needs a balancing act of making sure that information from all other waves is incorporated in the imputation of variables in each wave.

The following strategy within the sequential regression framework can be easily automated:

1. Use the propensity score for the response indicator for each variable with variables from all other waves and all other variables from the current wave as predictors. Suppose that Y_j is the variable from wave j to be imputed, and R_j is the missing data indicator. Let $X_{(-j)}$ denote the completed variables from all other waves and other variables from the wave in question. Let $\pi_j = Pr(R_j = 1|X_{(-j)})$ be the propensity score, and it is estimated, say, by fitting a logistic regression model.

2. Create H strata based on the estimated propensity score, $\widehat{\pi}_j$. Within each stratum use a regression model with Y_j as the dependent variable and $\widehat{\pi}_j$ as the predictor.

3. To further simplify, $X_{(-j)}$ may be decomposed into principal components to further reduce the dimensionality problem. Flexible models could be used for constructing the predictive distribution of Y_j given $\widehat{\pi}_j$ within each stratum.

The goal is to use as much information as possible and succinctly in the imputation process.

6.5   Weighting Methods

Weighting approach discussed in Chapter 2 can be extended for longitudinal data analysis. To be concrete suppose that, in a study with n subjects, U_i is a vector of baseline covariates, Y_{it} is response at time $t = 1, 2, \ldots, p$ for subject $i = 1, 2, \ldots, n$. Assume that U_i is fully observed but some subjects drop out of the study at various points during the study but never return (monotone pattern of missing data).

A generalized estimating equation (GEE) is a popular approach for analyzing longitudinal data which requires specification of the regression function relating the response variable vector $Y_i = (Y_{i1}, Y_{i2}, \ldots, Y_{ip})^T$ to a set of covariates, X_i (typically, U_i, some functions of time and their interactions), $\mu_i = g(X_i^T \beta)$ and a covariance matrix $V_i(\mu_i, \phi)$. A generalized estimating equation is of the form

$$S(\beta) = \sum_i^n D_i(X_i, \beta)^T V_i^{-1}(Y_i - \mu_i) = 0$$

where $D_i = \partial \mu_i / \partial \beta$ is the derivative of the mean function μ_i with respect to β. Note that D is $p \times k$ matrix where k is the dimension of the parameter β, V is $p \times p$ matrix, Y_i and μ_i are both $p \times 1$ vector. Thus, $S(\beta)$ is a $k \times 1$ vector giving k equations with k unknown parameters. The parameter ϕ also needs to be estimated and, typically, residuals are used to estimate ϕ. The popularity of this approach stems from the fact that even if V is misspecified, a consistent estimate of β is obtained. Thus, the modeling task basically reduces to finding a good regression model relating Y to X.

Now with missing data, one may be tempted to use the equation,

$$S_{obs}(\beta) = \sum_i^n D_{i,obs}(X_{i,obs}, \beta)^T V_{i,obs}^{-1}(Y_{i,obs} - \mu_{i,obs}) = 0$$

where each of the n components in the sum is based on the observed data on the corresponding subject. This strategy, however, results in a biased estimate of β unless the data are missing completely at random.

A weighted generalized estimating equation, introduces a weight for each subject to account for selection into a pattern determined by the index "*obs*." Specifically, let $R_{it} = 1$ if Y_{it} is observed and 0 otherwise. Assume that $R_{i1} = 1$ for all i (that is, the first wave is fully observed). Note that, under the monotone pattern of missing data, $R_{it} = 0$ implies that $R_{ij} = 0$ for all $j > t$.

Consider the fully observed subjects. They fall or are selected into this pattern, if $R_{it} = 1$ for all $t = 2, 3, \ldots, p$. Thus, the weight for subject i in this pattern is

$$w_{ip} = 1/Pr(R_{i2} = 1, R_{i3} = 1, \ldots, R_{ip} = 1 | U_i, Y_{i,(-p)})$$

where $Y_{i,(-p)}$ denotes all the Y values prior to the wave p.

Note that

$$Pr(R_{i2} = 1, R_{i3} = 1, \ldots, R_{ip} = 1 | U_i, Y_{i,(-p)}) =$$

$$Pr(R_{i2} = 1 | U_i, Y_{i,(-p)}) \times$$

$$Pr(R_{i3} = 1 | R_{i2} = 1, U_i, Y_{i,(-p)}) \times \ldots$$

$$Pr(R_{ip} = 1 | R_{i2} = 1, R_{i3} = 1, \ldots, R_{i,p-1} = 1, U_i, Y_{i,(-p)}). \qquad (6.3)$$

Note that $R_{ij} = 1$ implies that $R_{it} = 1$ for all $t < j$ and also missing at random implies that

$$Pr(R_{ij} = 1 | R_{i,j-1} = 1, U_i Y_{i,(-p)}) = Pr(R_{ij} = 1 | R_{i,j-1} = 1, U_i, Y_{i,(-j)}).$$

Thus the right hand side of equation (6.2) simplifies to

$$Pr(R_{i2} = 1 | U_i, Y_{i,(-2)}) \times$$

$$Pr(R_{i3} = 1 | R_{i2} = 1, U_i, Y_{i,(-3)}) \times \cdots$$

$$Pr(R_{ip} = 1 | R_{i,p-1} = 1, U_i, Y_{i,(-p)}).$$

Each conditional probability $\pi_{ij} = Pr(R_{ij} = 1 | R_{i,j-1} = 1, U_i, Y_{i,(-j)})$ can be estimated using a logistic regression model. These are the wave specific response propensities and $w_{ij} = 1/\hat{\pi}_{ij}$ are the wave specific nonresponse adjustment weights. The weights can be constructed for each wave using the response propensity method or other methods described in Chapter 2. In fact, one can also construct post-stratification weights for each wave to incorporate the population information.

The selection weight, for subject i, into the pattern with complete data is, therefore,

$$w_i = \prod_{t=1}^{p} w_{it}.$$

Similarly, the selection weight into the pattern with missing the last wave only is

$$w_i = \prod_{t=1}^{p-1} w_{it}.$$

In general, selection weight into a pattern of observing only the first j waves is

$$w_i = \prod_{t=1}^{t} w_{it}.$$

Let $w_{i,obs}$ be the selection weight for subject i for the observed pattern "obs" then the weighted generalized estimating equation is

$$S_{w,obs}(\beta) = \sum_{i} w_{i,obs} D_{i,obs}(X_{i,obs}, \beta)^T V_{i,obs}^{-1}(Y_{i,obs} - \mu_{i,obs}) = 0.$$

The weighting approach can be extended, at least conceptually, for the nonmonotone pattern of missing data. In the monotone case, the number of patterns of missing data is at most p. There can be a large number of patterns in the nonmonotone case (the worst case scenario is $2^{(p-1)} - 1$ patterns, assuming that wave one is fully observed. If wave one also has missing values then the number of possible patterns is $2^p - 1$). Thus, estimating the selection probability for a particular pattern for each subject becomes an arduous, if not impossible, task.

For $p = 4$, with wave one fully observed, there are seven possible patterns. As an example, consider a pattern $(R_{i2} = 1, R_{i3} = 0, R_{i4} = 1)$. This requires estimating the conditional probability of this pattern event given the observed data $(X_i, Y_{i1}, Y_{i2}, Y_{i4})$. With such complicated missing data patterns, it might be easier to multiply impute the missing values, then fit the GEE on each completed data and combine the point estimates and the standard errors to form a single inference.

Even with the monotone pattern, the weight construction requires specification of the missing data mechanisms (even under MAR). The simulation studies show that the results can be sensitive to misspecification of these mechanisms. Using poorly fitting models for estimating these response propensities can introduce bias and, hence, an approach that does not need exact specification of the missing data mechanism is preferable.

6.6 Binary Example

This chapter concludes with an example analysis of a binary longitudinal data with missing values. The American Changing Lives is a panel survey that began with a probability sample of 3,617 adults ages 25 and older in 1986. Subjects were reinterviewed in 1989, 1994, 2001/2 and 2011. Data collected includes a wide range of sociological, psychological, mental and physical health items. The goal of this particular analysis is to assess the trends in cognitive impairment over time and see the effect of socio-economic status and race on this trend. The socio-economic status is a four category variable with 1 coded as low, 2 as lower middle class, 3 as upper middle class and 4 as high. For this analysis, two race categories are (African-American and nonhispanic white). The outcome is a binary variable 1: moderate to severe impairment

Table 6.4: Summary statistics based on the observed ACL data: (number missing) and percent with cognitive impairment. Only nonmissing cases in Wave 1 included. No age restrictions applied.

Socio	Race	Wave				
Economic Status		1	2	3	4	5
Low	W ($n = 498$)	28.3	(95) 25.6	(113) 57.4	(93) 78.0	(155) 88.3
	AA ($n = 539$)	47.9	(77) 45.2	(124) 71.3	(130) 81.4	(163) 85.1
Lower middle	W ($n = 682$)	15.4	(102) 12.1	(117) 29.7	(109) 49.4	(229) 60.0
	AA ($n = 361$)	22.2	(80) 15.7	(98) 44.5	(101) 50.0	(112) 55.0
Upper middle	W ($n = 723$)	12.7	(76) 4.5	(71) 17.3	(93) 25.1	(216) 30.4
	AA ($n = 193$)	18.7	(51) 6.3	(45) 21.6	(65)33.6	(56) 33.6
High	W ($n = 302$)	5.3	(22) 2.1	(21) 8.5	(30) 16.5	(63) 22.2
	AA ($n = 63$)	14.3	(14) 2.0	(16) 17.0	(22) 24.4	(23) 27.5

and 0: little or no impairment. Death is coded as severe impairment in this illustrative analysis.

Table 6.4 provides descriptive statistics, sample size, the number of missing observations and the percentage of subjects with moderate to severe cognitive impairment for the full sample. The numbers are stratified by wave and the combination of socio-economic and race categories. For further analysis, the sample was restricted to subjects who were 50 years or younger in 1986 to avoid crowding of death (due to old age) as impairment in the latter period of the study. Thus, this restriction reduces confounding of the SES and race differences over the 25 year period due to death at old age.

All the missing values were multiply imputed using the sequential regression approach with 25 imputations, each requiring 20 iterations. Figure 6.4 provides multiple imputation least square regression lines relating the proportion impaired to time in decades for the eight groups. The figure shows that African-Americans generally have more impairment at the baseline than whites and also more positive trend in the prevalence of impairment across all socio-economic status categories. The lines are somewhat parallel for the higher levels of socio-economic status for both African-Americans and whites.

To assess the statistical significance of differences in the trends between these eight groups, a generalized estimating equation was used. The mean function on the logit scale included age, gender, seven dummy variables for the eight socio-economic × race groups, time coded as (year of the survey -1986)/10, and the interaction between them (a total of 18 parameters,

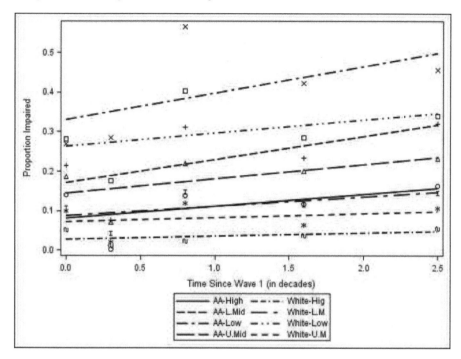

Figure 6.4: Least squares regression lines relating the proportion of subjects with impairment as a function of time (in decades) for the 8 SES-race group. Baseline age 50 years or younger.

including the intercept) as predictors. The covariance matrix was assumed to be unstructured (5×5).

A set of contrasts of interest is based on comparisons between African-Americans and whites within each socio-economic status. Table 6.5 provides the baseline odds ratio and change in the odds ratio over time for various comparison groups. The baseline odds ratio column shows that African-Americans have higher prevalence rate of impairment when compared to whites, across all four SES categories. The rate for African-Americans is a little over three times the rate for whites in the high SES category and almost 1 1/2 times higher in the low socio-economic status group. The rate change in the odds for African-Americans is larger than for the whites in all four groups (7% to 14% per decade). However, none of the rate of changes are statistically significant. High prevalence rate of impairment is maintained among the African-

Table 6.5: Comparison of intercepts and slopes between African-Americans and whites in each socio-economic status group. Baseline age 50 years or younger.

SES Status	Baseline OR (95% CI)	OR (per decade) (95% CI)
Low	1.45 (0.94,2.25)	1.14 (0.81, 1.59)
Lower middle	2.10 (1.40,3.14))	1.10 (0.84,1.44)
Upper middle	2.20 (1.46,3.33)	1.12 (0.84, 1.48)
High	3.28 (1.30, 8.26)	1.07 (0.57,2.02)

Americans through out this 25 year period, and perhaps have gotten worse compared to whites.

6.7 Bibliographic Note

A common method in longitudinal clinical trials is the LOCF (last observation carried forward) approach. A review of this and other early approaches can be found in Heyting, Tolboom and Essers (1992). Lavori, Dawsor and Shera (1995) proposed multiple imputation methods. The maximum likelihood approach using the EM-algorithm is discussed in Laird and Ware (1982) and Dempster, Rubin and Tsutakawa (1981). Little (1995) provides a lucid discussion of modeling issues with drop outs in the longitudinal studies. Little and Rubin (2002) also provide extensive discussion of the maximum likelihood estimation for longitudinal data. Siddiqui and Ali (1998) and Mallinckdrot et al (2001, 2003) compare various approaches for analyzing incomplete data.

Diggle, Heagerty, Liang and Zeger (2002), Verbeke and Molenberghs (2000) and Molenberghs and Verbeke (2005) provide comprehensive coverage of issues in the analysis of longitudinal data including an extensive discussion of missing data. Another good read is the report by National Research Council (2010).

The inverse probability weighting and general semiparametric approaches for analyzing repeated measures data with missing values are discussed in

Robins, Rotnitzky and Zhao (1995). See also Davidian, Tsiatis and Leon (2005). Preisser, Lohman and Rathouz (2002) assess the performance of weighted estimating equation, especially with respect to mispecification of the missing data mechanism. The primary reference for Bup-Nx and clonidine comparison trial is Ling et al (2005) and for the American Changing Lives (ACL) is the ICPSR archive, House (2014).

6.8 Exercises

1. Consider a two-treatment randomized clinical trial where baseline measures $y_{oi}, i = 1, 2, \ldots, 2n$ of an outcome variable, Y, was obtained on a random sample of $2n$ individuals. The subjects were randomized either to a treatment $(T = 1)$ or the control $(T = 0)$ group. Only r_0 (r_1) subjects showed up in the treatment (control) at the follow-up where the same outcome measure was obtained on $r = r_0 + r_1$ subjects, $y_{1i}, i = 1, 2, \ldots, r$. The parameter of interest is the population mean difference at the follow-up, $\theta = E(Y_1|T = 1) - E(Y_1|T = 0)$. Here are the possible estimates.

   (a) Estimate based on completers,

   $$\widehat{\theta}_1 = \bar{y}_{11} - \bar{y}_{10} = \sum_i^{r_1} y_{1i}/r_1 - \sum_j^{r_0} y_{1j}/r_0.$$

   (b) Create H strata based on y_0. Let n_{1h} and n_{0h} be the treatment and control sample sizes, respectively, in stratum $h = 1, 2, \ldots, H$, with the corresponding numbers of respondents, r_{1h} and r_{0h}. Let $\bar{y}_{11h}$ $(\bar{y}_{10h})$ be the mean of the outcome at the follow-up for the treatment (control) respondents in stratum h. Define,

   $$\widehat{\theta}_2 = \sum_h n_{1h}\bar{y}_{11h}/\sum_h n_{1h} - \sum_h n_{0h}\bar{y}_{10h}/\sum_h n_{0h}.$$

   (c) Suppose that it is reasonable to assume $y_1|y_o \sim N(\beta_o + \beta_1 T + \beta_2 y_0 + \beta_3 T \times y_0, \sigma^2)$. The third estimate, $\widehat{\theta}_3$, is the multiply imputed estimate based M imputation created from the model assuming the prior, $Pr(\beta_o, \beta_1, \beta_2, \beta_3, \sigma) \propto \sigma^{-1}$.

Discuss the pros and cons of each estimate. Some analysts may not be comfortable in explicit imputation of the outcome values and for that reason may prefer estimates $\widehat{\theta}_1$ or $\widehat{\theta}_2$. What arguments can be used to convince that both estimates involve implicit imputation?

2. Perform your own multiple imputation of missing values in the data discussed in Section 6.3. Fit the following *autoregressive* model,

$$y_{it} = \beta_{oi} + \beta_{1i}y_{i,t-1} + \beta_{2i}y_{i,t-2} + \epsilon_{it},$$

where $\epsilon_i \sim N(0, \sigma^2)$ and $\beta_i = (\beta_{oi}, \beta_{1i}, \beta_{2i})$ has a trivariate normal distribution with mean $\beta = (\beta_o, \beta_1, \beta_2)$ and a covariance matrix Ω, on each completed data set. Assume that βs and ϵs are independent. Construct multiple imputation inferences for β. How will you assess the fit of this model and compare it with the models discussed in Section 6.3?

3. The American Changing Lives (ACL) study discussed in Section 6.6 has several other variables. Choose another outcome variable and repeat the analysis.

7

Nonignorable Missing Data Mechanisms

7.1 Modeling Framework

So far, the missing data mechanism was assumed to be ignorable where both, multiple imputation and the maximum likelihood approaches, do not require exact specification of the missing data mechanism. The weighting approach, however, does require specification of the missing data mechanism. When the data are not missing at random then all three approaches require specification of the missing data mechanism, which involves parameters as well as unobserved values.

To be concrete, consider a bivariate data on n subjects with X, fully observed, and Y observed for subjects $i = 1, 2, \ldots, r$, and missing for $i = r + 1, r + 2, \ldots, n$. Under the missing not at random mechanism (MNAR), the specification of the joint distribution of (Y, R) given X is needed, where $R = 1$ for the first r subjects and 0 for the remaining $n - r$ subjects. Consider two ways to specify the joint distribution as discussed in Chapter 1.

A selection model specification defines the joint distribution as,

$$Pr(Y, R|X, \theta, \psi) = Pr(Y|X, \theta)Pr(R|Y, X, \psi).$$

As an example, consider a normal linear regression model for Y given X, $Y|X, \theta \sim N(\theta_o + \theta_1 X, \sigma^2)$, and a probit model

$$\pi(X, Y|\psi) = Pr(R = 1|Y, X, \psi) = \Phi(\psi_o + \psi_1 X + \psi_2 Y)$$

for R where $\Phi(u) = Pr(Z \leq u)$ is the cumulative distribution function of the standard normal random variable.

The likelihood based approach requires maximization of,

$$L(\theta, \psi|Y_{obs}, X, R) \propto \int Pr(Y|X, \theta)Pr(R|Y, X, \psi)dY_{mis}.$$

The probit model shows that there is no information in the data to estimate ψ_2 as all missing Ys correspond to $R = 0$. That is, the likelihood will be "flat" with respect to ψ_2. Thus, to proceed further, ψ_2 needs to be specified and then the remaining parameters can be obtained by maximizing the likelihood.

The mixture model specification defines the joint distribution as

$$Pr(R, Y | X, \alpha, \beta) = Pr(Y | X, R, \alpha) Pr(R | X, \beta).$$

To be concrete, consider $Y | X, R = 1, \alpha_1 = (\alpha_{1o}, \alpha_{11}), \sigma_1 \sim N(\alpha_{1o} + \alpha_{11} X, \sigma_1^2)$ and $Y | R = 0, \alpha_o = (\alpha_{oo}, \alpha_{o1}), \sigma \sim N(\alpha_{oo} + \alpha_{o1} X, \sigma_o^2)$ and $Pr(R = 1 | X, \beta) = \pi(X | \beta) = \Phi(\beta_o + \beta_1 X)$. Under this specification, the observed data provides information to estimate $\beta_o, \beta_1, \alpha_{1o}, \alpha_{11}$ and σ_1^2. There is no information to estimate α_{oo}, α_{o1} and σ_o^2. The complete data model for Y given X is a mixture of the normal distributions,

$$\pi(X | \beta) N(\alpha_{1o} + \alpha_{11} X, \sigma_1^2) + (1 - \pi(X | \beta)) N(\alpha_{oo} + \alpha_{o1} X, \sigma_o^2),$$

which is different than what is assumed under the selection model. However, the validity of the complete data model under either framework (selection or pattern-mixture) cannot be empirically assessed (using any diagnostic procedures, for example) because of the lack of information about the missing Ys.

7.2   EM-Algorithm

Before discussing the construction of inference about the parameters using the selection model framework, the EM-algorithm is described. This is an easy to implement algorithm for obtaining maximum likelihood estimates. The advantage is that the algorithm does not involve computing first or second derivatives of the log-likelihood function, as opposed to, say, Newton-Raphson or its variants. This algorithm is closely connected to the idea of imputation, except that instead of imputing actual variables, some statistics (aggregates) are imputed (through their expected values).

Briefly, the maximum likelihood estimates are the functions of "sufficient statistics," which are the only functions of the data that are needed to compute the maximum likelihood estimates. Technically, suppose that $y_1, y_2, \ldots, y_n$ is a random sample from the distribution with the density function $f(y|\theta)$, where

θ is the unknown parameter and the object of inference based on the sample. Suppose $T = T(y_1, y_2, \ldots, y_n)$ is a statistic (it could be a vector). The statistic T is called sufficient for θ, if

$$Pr(y_1, y_2, \ldots, y_n | \theta, T = t) = g(y_1, y_2, \ldots, y_n)$$

where g does not involve θ. That is, given T there is no additional "information" about θ in the rest of the data.

As an example, suppose that $y_1, y_2, \ldots, y_n$ is a random sample from a Bernoulli distribution with parameter θ then $T = \sum_i y_i$ is a sufficient statistic. Similarly, if $y_1, y_2, \ldots, y_n$ is a random sample from a normal distribution with mean μ and variance σ^2, sufficient statistics are the sample mean and variance or, equivalently, the sample total $\sum_i y_i$ and sample sum of squares $\sum_i y_i^2$, the building blocks for computing the sample mean and variance.

The EM-algorithm is an iterative algorithm involving two steps. The E-step of the EM-algorithm uses a starting value of the parameters, estimates the missing portions of the sufficient statistics by their expectation (conditional on the observed data and the current value of the parameters). The M-step uses the observed and estimated sufficient statistics to obtain revised parameter estimates. The iterations continue until the parameter estimates converge (stabilize) or does not change up to a certain decimal place.

A simple, though trivial, example is useful to illustrate the underlying basic idea. Suppose that $x_1, x_2, \ldots, x_n$ is a random sample of size n from a normal distribution with unknown mean μ and variance σ^2. The sufficient statistics for estimating μ is the sum $S_1 = \sum_i x_i$ because $\hat{\mu} = S_1/n = \bar{x}$, the sample mean. An additional sufficient statistic for estimating variance is $S_2 = \sum_i x_i^2$ because

$$\hat{\sigma}^2 = \sum_i (x_i - \bar{x})^2/n = S_2/n - \bar{x}^2 = (S_2 - S_1^2/n)/n.$$

Suppose that only the first r values are observed and the remaining $n - r$ are missing and the missing data mechanism is ignorable. Obviously, the correct estimates for the mean μ in the sample mean of the observed values $\bar{x}_R = \sum_i^r x_i/r$ and the estimate of the variance, σ^2, is the sample variance (with the divisor r, rather than $r - 1$), $s_R^2 = \sum_i (x_i - \bar{x}_R)^2/r$.

Nevertheless, consider using the EM-algorithm to obtain the maximum likelihood estimates. The missing portions of the two sufficient statistics are the unobserved sum, $\sum_{i=r+1}^n x_i$ and the unobserved sum of squares, $\sum_{i=r+1}^n x_i^2$.

Suppose μ_o and σ_o^2 are the initial values. The E-step is to estimate the missing portions of the sufficient statistics by taking their expectations conditional on the parameters and the observed values, which are $E(\sum_{i=r+1}^{n} x_i) = (n-r)\mu_o$ and $E(\sum_{i=r+1}^{n} x_i^2) = (n-r)(\sigma_o^2 + \mu_o^2)$.

The M-step obtains revised estimates using the estimated sufficient statistics,

$$\mu_1 = \frac{\sum_i^r x_i + (n-r)\mu_o}{n},$$

and

$$\sigma_1^2 = \frac{\sum_i^r x_i^2 + (n-r)(\sigma_o^2 + \mu_o^2)}{n} - \mu_1^2.$$

Revised estimated sufficient statistics are $(n-r)\mu_1$ and $(n-r)(\sigma_1^2 + \mu_1^2)$ which are then used to obtain μ_2 and σ_2^2. This continues until the successive values of μ and σ are the same up to certain decimal places.

At the convergence $\mu_t = \mu_{t-1}$ and $\sigma_t^2 = \sigma_{t-1}^2$ (the equality is up to certain decimal places) which implies that

$$\mu_t = \frac{\sum_i^r x_i + (n-r)\mu_t}{n}$$

and simplifies to $\mu_t = \bar{x}_R = \sum_i^r x_i / r$. Similarly,

$$\sigma_t^2 = \frac{\sum_i^r x_i^2 + (n-r)(\sigma_t^2 + \mu_t^2)}{n} - \mu_t^2$$

and simplifies to

$$\sigma_t^2 = \sum_i^r x_i^2 / r - \mu_t^2 = s_R^2 = \sum_i (x_i - \bar{x}_R)^2 / r.$$

7.3   Inference under Selection Model

The EM algorithm can be used to obtain maximum likelihood estimates of the parameters in the selection model framework for a given value of ψ_2. A convenient representation of the probit model is through a latent variable U having a normal distribution with mean $\psi_o + \psi_1 X + \psi_2 Y$ and variance 1 and then defining $R = 1$ if $U > 0$ and 0 otherwise.

Thus, the two regression models defining the selection model framework are $Y|X \sim N(\theta_o + \theta_1 X, \sigma^2)$ and $U|Y, X \sim N(\psi_o + \psi_1 X + \psi_2 Y, 1)$ where all U are missing, except it is known that when $R = 1$, $U > 0$ and when $R = 0$, $U \leq 0$. Also Y is missing when $U \leq 0$. The probit model is popular because the missing data framework can be used by treating U as missing with partial information provided by R. There are many other applications (such as factor analysis, atructural equation models, random effects models, etc.) where "augmented data" is created with observables and unobservables and then EM-algorithm framework is used to obtain the maximum likelihood estimates as a function of observed variables.

To identify complete data sufficient statistics, assume that all Y and U are observed. Thus, based on the complete data, fitting the first regression model results in the following estimates,

$$\widehat{\theta}_1 = \frac{\sum_i^n Y_i (X_i - \bar{X})}{\sum_i^n (X_i - \bar{X})^2},$$

$$\widehat{\theta}_o = \frac{1}{n} \sum_i^n Y_i - \widehat{\theta}_1 \bar{X}$$

and

$$\widehat{\sigma}^2 = \frac{1}{n} \sum_i^n (Y_i - \widehat{\theta}_o - \widehat{\theta}_1 X_i)^2.$$

Define $U_i^* = U_i - \psi_2 Y_i$. The second model is then a regression of U^* on X and thus,

$$\widehat{\psi}_1 = \frac{\sum_i^n U_i^* (X_i - \bar{X})}{\sum_i^n (X_i - \bar{X})^2}$$

and

$$\widehat{\psi}_o = \frac{1}{n} \sum_i^n U_i^* - \widehat{\psi}_1 \bar{X}.$$

Consider now the unknown pieces in the above estimates of the parameters, $\lambda = (\theta_o, \theta_1, \sigma^2, \psi_o, \psi_1)$. The three unknown pieces are U for both $R = 1$ and $R = 0$; Y and Y^2 for $R = 0$. The E-step involves computing the expectation of the unknown quantities conditional on the observed data and the current value of the parameter.

The expectation of U when $R = 1$ is $E(U|Y, X, U > 0, \lambda)$ and when $R = 0$, it is $E(U|X, U \leq 0, \lambda)$. The expectation of Y and Y^2 when $R = 0$ are $E(Y|X, U \leq 0, \lambda)$ and $E(Y^2|X, U \leq 0, \lambda)$, respectively. All these expectations are easy to calculate based on the truncated normal and truncated bivariate normal distributions.

Note that

$$E(U|X) = E(E(U|X,Y)|X) = \psi_o + \psi_1 X + \psi_2(\theta_o + \theta_1 X)$$
$$= (\psi_o + \psi_2\theta_o) + (\psi_1 + \psi_2\theta_1)X = \psi_o^* + \psi_1^* X.$$

Also, $Var(U|X) = 1 + \psi_2^2 Var(Y|X) = 1 + \psi_2^2\sigma^2$ and $Cov(Y,U|X) = \psi_2\sigma^2$.

Thus, the joint distribution of (Y,U) is bivariate normal with mean,

$$\begin{bmatrix} Y \\ U \end{bmatrix} \sim N\left(\begin{bmatrix} \mu_{1X} = \theta_o + \theta_1 X \\ \mu_{2X} = \psi_o^* + \psi_1^* X \end{bmatrix}, \begin{bmatrix} \sigma^2 & \rho\sigma\sigma_1 \\ \rho\sigma\sigma_1 & \sigma_1^2 \end{bmatrix} \right),$$

where $\rho = \psi_2\sigma/\sqrt{1 + \psi_2^2\sigma^2}$ and $\sigma_1^2 = 1 + \psi_2^2\sigma^2$.

Based on the above joint distribution, the standard truncated normal and truncated bivariate normal distributions imply the following:

1. $E(U|X, U \leq 0, \lambda)$ is given by

$$\mu_{2X} + \sigma_1 M(-\mu_{2X}/\sigma_1)$$

where $M(z) = \phi(z)/\Phi(z)$ is the Mill's ratio, the ratio of the density to the cumulative distribution function of the standard normal distribution, evaluated at z.

2. For computing $E(U|Y, X, U > 0, \lambda)$, first note that $U|Y, X, \lambda$ is normal with mean

$$\mu_* = \mu_{2X} + \rho\sigma_1(Y - \mu_{1X})/\sigma$$

and variance $\sigma_*^2 = \sigma^2(1 - \rho^2)$. Incorporation of truncation obtains

$$E(U|YX, U > 0, \lambda) = \mu_* + \sigma_* M(\mu_*/\sigma_*).$$

3. $E(Y|X, U \leq 0, \lambda)$ is given by

$$\mu_{1X} - \rho\sigma M(-\mu_{2X}/\sigma_1).$$

4. $E(Y^2|X, U \leq 0, \lambda)$ is given by

$$Var(Y|X, U > 0, \lambda) + E^2(Y|X, U > 0, \lambda)$$
$$= \mu_{1X}^2 + \sigma^2 + \sigma M(-\mu_{2X}/\sigma_1)(2\mu_{1X} - \rho^2\sigma\mu_{2X}/\sigma_1).$$

The above setup can be extended for a multivariate X with no missing values. For instance, the complete data regression models are multiple linear regression and define $\mu_{1X} = \theta_o + \theta_1^T X$ and $\mu_{2X} = \psi_o + \psi_1^T X$. There are no readily available software packages for implementing the EM algorithm. Computer programs have to be developed by the user using macro facilities (such as IML routines in SAS).

7.4 Inference under Mixture Model

Multiple imputation is easy to implement under the mixture model framework as it can be viewed as a departure from the imputation under MAR. As discussed earlier, the observed data does not provide information about $(\alpha_{oo}, \alpha_{o1})$ and σ_o^2. Define $\alpha_{oo} = (1+a)\alpha_{1o}$, $\alpha_{o1} = (1+a)\alpha_{11}$ and $\sigma_o = (1+a)\sigma_1$ where $a = 0$ represents MAR and various choices of a (like ψ_2 in the selection model) specify departures from MAR or the extent of nonignorability of the missing data mechanism.

As discussed in Chapter 3, suppose that σ_1^*, α_{1o}^*, α_{11}^* and Y_i^*, $i = r+1, r+2, \ldots n$ are the drawn values of the parameters and the missing values based on the observed data assuming MAR. The imputed values for the nonrespondents under the mixture model are $Y_i^{(o)} = (1 + a)Y_i^*$, $i = r + 1, r + 2, \ldots, n$.

This approach is easy to implement by multiplying the imputed values under MAR by a fixed number $(1+a)$. Sensitivity of inferences to assumption about the missing data mechanism can be assessed by varying the values of a. An alternative option is to randomly perturb the imputed values for the nonrespondents,

$$Y_i^{(o)} = (1 + u_i)Y_i^*, i = r + 1, r + 2, \ldots, n$$

where u_i are independent uniform random numbers between 0 and $2a$.

7.5 Example

Consider the data from the randomized trial comparing Bup-Nx and clonidine for detoxification that was analyzed in Chapter 6. The focus for this example analysis is on the outcome, Y, the value of the visual analog score (VAS) on day 14. The covariates are age, sex (female), race (white), treatment (Bup-Nx) indicator and the baseline value of VAS. The primary parameter of interest is the regression coefficient for the treatment indicator variable in the following regression model,

$$y = \beta_o + \beta_1 Age + \beta_2 Female + \beta_3 White + \beta_4 BaseVAS + \beta_5 Treat + \epsilon.$$

The imputed values for Y obtained under MAR assumption in Chapter 6 can be perturbed to various extents for either or both treatment groups.

Table 7.1: Estimated treatment effect (SE) on Day 14 visual analog score (VAS) adjusted for age, gender, race and baseline VAS

Perturbation	Both Groups	Bup-Nx Only	Clonidine Only
0% (MAR)	-14.32 (8.78)	-14.32 (8.78)	-14.32 (8.78)
2.5%	-14.89 (9.02)	-14.13 (8.82)	-15.07 (8.96)
5%	-15.40 (9.11)	-13.95 (8.82)	-15.75 (9.10)
10%	-16.41 (9.55)	-13.63 (8.86)	-17.25 (9.61)
20%	-18.60 (10.42)	-12.96 (9.07)	-20.06 (10.37)
30%	-21.07 (11.74)	-12.19 (9.03)	-23.27 (11.08)

Table 7.1 summarizes the results of perturbing the imputed values by 2.5%, 5%, 10%, 20% and 30% larger than those obtained under MAR. The second column is for perturbing both treatment groups, the third is for perturbing only the treatment (Bup-Nx) group and the last column is for perturbing only the clonidine group. It is clear from the results that even under the conservative scenario where the nonresponding subjects only in the treatment group are expected to have 30% higher scores than the responding subjects, Bup-Nx still has a slight edge over the clonidine group.

7.6   Practical Considerations

As emphasized in Chapter 1, both MAR and MNAR mechanisms are empirically unverifiable and the point of the short story in Section 1.6 is to underscore the difficulties in the specification of the missing data mechanism. However, MNAR modeling framework requires specification of the missing data mechanism where some parameters are not estimable without imposing restrictions or additional information.

From the practical point of view, one needs to consider strategies for making the MAR mechanism plausible. This implies collecting relevant covariate information related to the outcome and then including them in the modeling framework. For example, to handle missing values in the income variables, contextual measures such neighborhood level property values, type of housing structure, industry and occupation codes and many other predictors of income may have to be collected regardless whether they are to be used in the substantive model.

As another example, information relevant to drop out such as adverse events, toxicity, side effects, etc. have to be collected in a randomized clinical trial. These have to be included in the imputation process to make MAR plausible. Thus, preparation for the analysis of incomplete data needs to be considered at the design stage and plan for relevant additional data collection.

Instead of choosing specific nonignorable missing data mechanisms, it may be useful to perform sensitivity analysis by perturbing the imputed values obtained under the MAR assumption. The pattern-mixture modeling framework makes the perturbation mechanism relatively easy to implement. The goal is to assess the robustness of findings with respect to the departures from the MAR mechanism. One can be more confident with respect to statistically significant findings, if a "large" departure from the MAR mechanism is needed to eliminate the statistical significance. Of course, it is a nontrivial task to exactly define a large departure.

7.7 Bibliographic Note

The selection model formulation is based on Heckman (1976) where it was used to model the selection of women into the labor force. Lillard, Smith and Welch (1986) is an application of this approach to handle missing income values in the current population survey. Little (1985) and Little and Rubin (1987) show that this approach is highly sensitive to the stated assumptions.

Rubin (1977) proposed the simple mixture model framework for a scalar variable based on normal distribution. Glynn, Laird and Rubin (1986) report on several simulation studies and consider extensions with covariates. Little (1993, 1994, 1995) is a series of influential articles advocating the pattern-mixture model for handling nonignorable missing data mechanism in a variety of contexts.

Pregibon (1977), Little (1982), Nordheim (1984), Baker and Laird (1988) and Stasny (1986) are some references that consider nonignorable missing data models for categorical data. The models consider the joint cross-classification of substantive and missing data indicator variables. The parameters that cannot be estimated are either fixed at various values or handled through a prior distribution.

Considerable literature exist on selection and mixture modelling framework for handling missing values in the longitudinal analysis context. Some of the books listed in Chapter 6 cover these topics extensively. Additional references from a Bayesian perspective include Kaciroti et al (2006, 2008, 2009, 2014).

7.8   Exercises

1. Consider the data given in Table 4.1. Perform your own multiple imputation analysis. Suppose one suspects that mothers who are actually smoking are more likely to be missing smoking status. Perform sensitivity analysis to assess the impact of this suspicion on the association between maternal smoking and child wheeze status.

2. In the previous problem, suppose one suspects that mothers who are actually smoking and have wheezing kids are more likely to be missing smoking status. Carry out sensitivity analysis for assessing the impact of this assumption on the association between maternal smoking and child wheeze status.

3. In a randomized study, the goal was to compare the distribution of the outcome variable, Y, between two groups $G = 1$ and $G = 2$. Suppose X is a covariate with no missing values. The outcome had some missing values. The investigator suspects that the missing data mechanism may be nonignorable and would like to know how different the respondents and nonrespondents have to be for the differences between the two groups to be statistically not significant. Develop analysis plans to answer investigators questions.

4. Refer to problem 7 in Chapter 5. Suppose that censored subjects have a different distribution for the response variable. That is, for subjects with observed failure times, the distribution is $y \sim N(\beta_o + \beta_1(x_1 - \bar{x}_1) + \beta_2(x_2 - \bar{x}_2), \sigma^2)$ but for the censored subjects, the distribution is $y \sim N((1+a)\beta_o + \beta_1(x_1 - \bar{x}_1) + \beta_2(x_2 - \bar{x}_2), (1+b)^2\sigma^2)$ where a and b are fixed constants. Perform multiple imputation analysis for inferring about the probability of survival beyond eight days under these model assumptions for various choices of a and b. Develop appropriate graphical or numerical displays.

8

Other Applications

Missing data framework and, especially, the multiple imputation method can be applied to other situations. This chapter discusses five applications: measurement error, combining information from multiple data sources, Bayesian inference for finite population, causal inference and statistical disclosure control. Each situation can be handled using standard multiple imputation software. The same combining rules described so far can be applied in some applications but for others, new combining rules have to be used.

8.1 Measurement Error

Suppose that Y is an outcome measure and X is a covariate, and the goal is to fit a regression model, $Y = g(X, \theta) + \epsilon$, where g is a known function of X with some unknown parameters θ, ϵ follows a known distribution and may involve θ, X and some additional parameters ϕ. The covariate X is expensive to measure but a surrogate or less expensive substitute, W, is available. The main study thus measures Y and W and a substudy may be conducted to measure X, Y or W. Figure 8.1 provides three example scenarios, with the main and substudies.

In scenario (a), the main study collects information on the pair (Y, W) and a substudy collects data on all three variables (Y, W, X). Here the goal is to append (or vertically concatenate) the substudy data set to the main study data set and then multiply impute the missing X in the main study. The prediction model needed is $f(X|Y, W)$. Here the missing at random assumption implies that this conditional distribution is the same for both main study and the substudy.

After the imputation of missing Xs in the main study, the analyst may use observed Y from both studies, observed X from the substudy and the

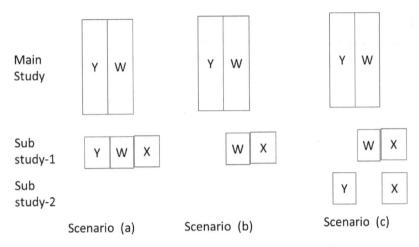

Figure 8.1: Measurement error study scenarios

multiply imputed X from the main study for the main regression analysis. An alternative option is to discard the substudy after the imputation and just use the main study with observed Y and multiply imputed X. The latter option might be more appropriate, if the substudy was specifically designed using a laboratory setting to accurately estimate the conditional density function $f(X|Y,W)$.

As an example of scenario (a), consider a situation where a large survey collects information about various diseases by asking questions "Did a Doctor or other health professional ever tell you that you have disease, D"? This may not accurately reflect the actual disease status, if the subject has not been seen by a doctor or other health professional. Suppose that another study is available that not only asks the same set of questions but also performs a medical examination, laboratory analysis of specimens collected or abstraction of information from medical records. Schenker et al (2010) apply the multiple imputation methodology by considering the "main study" as the National Health Interview Survey (NHIS) and the "substudy" as the National Health and Nutritional Examination Survey (NHANES).

Scenario (b) is, perhaps, the most common measurement error setup where the main study collects information on Y and W whereas the substudy collects data on W and X. When the data from the two studies are vertically concatenated, it produces a "file matching" pattern of missing data described in Figure 1.1. Imputation of missing values in X in the main study and Y

in the substudy needs to be carried without measuring Y and X on any subjects. In other words, there is no information in the data about the partial correlation between Y and X conditional on W.

Since both Y and X have to be imputed in the concatenated data set, one needs the predictive distribution, $f(X, Y|W)$, which can be decomposed into $f(Y|X, W) \times f(X|W)$. It is reasonable to assume that $f(Y|X, W) = f(Y|X)$ because, if X were available then W has no information about the substantive model of interest $f(Y|X)$. The imputation process needs two models, $f(Y|X)$ and $f(X|W)$. Unfortunately, there is no empirical basis for determining the appropriateness of $f(Y|X)$ and is purely driven by the substantive theory. The substudy provides information about $f(X|W)$. The missing at random assumption implies that $f(X|W)$ is the same for both the main study the and the substudy.

To be concrete, suppose that Y, X and W are continuous and normally distributed and the goal is to fit a regression model, $Y = \beta_o + \beta_1 X + \epsilon$ with $\epsilon \sim N(0, \sigma^2)$. Further assume that $X = \mu_X(W) + \eta = \alpha_o + \alpha_1 W + \eta$ where $\eta \sim N(0, \tau^2)$.

For imputation, one needs $f(X|Y, W)$ which is given by,

$$f(X|Y, W) = \frac{f(X, Y|W)}{f(Y|W)} = \frac{f(Y|X, W) f(X|W)}{f(Y|W)} = \frac{f(Y|X) f(X|W)}{f(Y|W)}$$

$$\propto f(Y|X) f(X|W) \propto \exp[-\{(Y - \beta_o - \beta_1 X)^2 / \sigma^2 + (X - \alpha_o - \alpha_1 W)^2 / \tau^2\}]$$

$$\propto \exp[-\{(X - (Y - \beta_o)/\beta_1)^2 / (\sigma^2/\beta_1^2) + (X - \alpha_o - \alpha_1 W)^2 / \tau^2\}].$$

Combining the two normal densities obtains,

$$X|Y, W \sim N(\mu_X(Y, W), \psi^2) \tag{8.1}$$

where $\psi^2 = [\beta_1^2 / \sigma^2 + 1/\tau^2]^{-1}$ and

$$\mu_X(Y, W) = \psi^2 [\beta_1 (Y - \beta_o)/\sigma^2 + (\alpha_o + \alpha_1 W)/\tau^2].$$

Note that $Y|W \sim N(\mu_Y^*(W) = \beta_o^* + \beta_1^* W, \sigma_*^2)$ where

$$\beta_o^* = \beta_o + \beta_1 \alpha_o,$$

$$\beta_1^* = \beta_1 \alpha_1,$$

$$\sigma_*^2 = \beta_1^2 \tau^2 + \sigma^2. \tag{8.2}$$

Assuming a noninformative prior distribution for the parameters $\omega = (\beta_o^*, \beta_1^*, \alpha_o, \alpha_1, \sigma_*, \tau)$, $\pi(\omega) \propto \sigma_*^{-1}\tau^{-1}$, It is straightforward to draw values of α_0, α_1 and τ^2 based on the regression analysis of X on W (from the substudy). Next, it is easy to draw values of β_o^*, β_1^* and σ_*^2 based on the regression analysis Y on W (from the main study).

Using the drawn values, of $(\alpha_o, \alpha_1, \tau^2, \beta_o^*, \beta_1^*, \sigma_*^2)$, one can use equations (8.2), to solve for β_o, β_1 and σ and then use (8.1) to draw values from the conditional predictive distribution of X, given Y and W, in the main study. Note that, to implement this approach, only the sufficient statistics are needed from the substudy and not the actual data set. This is useful, when a laboratory performs substudies publishes "measurement error" analysis results.

An alternative approach is to use a sequential regression approach where the nonestimable parameters are set to some fixed values. In this situation, there is no information in the data to estimate the partial correlation coefficient between Y and X conditional on W. If the concatenated data from the two studies were used as input into a program (like IVEware), it can be shown that the multiple imputation estimate of this partial correlation coefficient will be zero as the number of imputations tends to ∞.

A more general strategy to generate imputation for prespecified values of the partial correlation coefficient, $\rho_{YX.W}$ (this partial correlation coefficient is the simple correlation coefficient between the residuals from the regression of Y on W and the residuals from the regression of X on W), is as follows. Suppose that, based on the main study, it is reasonable to assume that $Y|W \sim N(\mu_Y(W), \sigma^2)$ and based on the substudy, $X|W \sim N(\mu_X(W), \tau^2)$. Both models can be determined through separate empirical analysis of the main study and substudy. Further, assume a bivariate normal distribution for $(X, Y)|W$ which implies that

$$X|Y, W \sim N[\mu_X(W) + \rho\tau/\sigma(Y - \mu_Y(W)), \tau^2(1 - \rho^2)]$$

where ρ is a fixed partial correlation coefficient between X and Y given W.

As before, the draws of the unknown parameters in $(\mu_X(W), \tau^2)$ based on the substudy, and the draws of $(\mu_Y(W), \sigma^2)$ from the main study, and a fixed value of ρ could be used to generate multiple imputations of X in the main study. A more general setup is to make σ and τ also depend upon W (for example, heteroscedastic error distribution).

A misclassification problem occurs when all three, Y, X and W, are binary variables. The assumption $f(Y|X,W) = f(Y|X)$ implies that the overall odds ratio between Y and X is equal to the two conditional odds ratios for $W = 0$ and $W = 1$.

Scenario (c) occurs when an additional substudy is conducted where (Y, X) is collected on a different set of subjects. It may be a result of a designed experiment where subjects representing a broad distribution of Y are recruited and X is measured on them. In a trivariate normal case with (Y, X, W), now the partial correlation coefficient between Y and X given W can be estimated. A standard sequential regression approach or multivariate normal approach can be used to impute the missing values of X in the main study as well as missing Y and W in the concatenated data set.

8.2 Combining Information from Multiple Data Sources

Increasingly, it is difficult to find one data source that can provide information to address complex substantive problems. The missing data framework can be exploited by vertically (or horizontally) concatenating multiple data sources resulting in a "jig saw" puzzle type data structure and then filling-in the missing values. To be concrete, suppose that a study collects information about diseases Y and their biological risk factors X and another study collects information about the same set of diseases Y but behavioral risk factors Z. There is also an omnibus survey that collects prevalence of biological and behavioral risk factors X and Z. Assume that all three studies are representing similar populations (the results from each study are "transportable" to the other two studies). A vertically concatenated data set then allows the estimation of the joint distribution (Y, X, Z) under the assumption that the three bivariate distributions are sufficient to define the trivariate distribution. One can expand on this notion by assembling a large number of data sets collecting different domains from various surveys, clinical studies, epidemiological investigations to expand the possibilities of the analysis that are impossible using individual data sets. For example, a data archival entity can explore the data sets in its archive, harmonize the variables through recoding, concatenate and create multiply imputed data sets. Such multiply imputed data sets can expand the research infrastructure much beyond the individual data sets in the repository.

It is possible that when several data sets are vertically concatenated, some parameters are not estimable. Consider a situation where a study collects information about Y and X and another study collects information about Y and Z. The third study collects information about Y_o, X_o and Z_o, a collection of a subset of variables from Y, X and Z, respectively. A small study may be mounted to collect data on the remaining portions of Y, X and Z, or the nonestimable parameters can be obtained based on the published literature.

The strategy of leveraging information from several studies through missing data framework and, especially, through multiple imputation is going to be critical for scientific research given that primary data collection is becoming expensive and a wealth of information from administrative data and other nonsurvey data will become more available. However, such pooling of information from multiple sources also increases concerns about privacy and confidentiality. Multiple imputation framework can also be used to address these concerns, as discussed in Section 8.4.

8.3 Bayesian Inference from Finite Population

In a finite population inference, the goal is to estimate the characteristics, Q, (such as mean or variance) of the collection of measurable values $(Y_1, Y_2, \ldots, Y_N)$ based on a sample $(Y_{i_1}, Y_{i_2}, \ldots, Y_{i_n})$ where $s = (i_1, i_2, \ldots, i_n)$ is a sample or a subset of $P = (1, 2, \ldots, N)$. The sample is a deliberate choice (usually using some random mechanism) made by the investigator. Additional information $(Z_1, Z_2, \ldots, Z_N)$ may have been used in the sampling process.

Without loss of any generality, the observed subset (sampled values) may be listed first and the nonsampled values may be listed next and thus creating a standard missing data setup where (Z, Y) are observed on the sampled subjects and only Z is observed on the nonsampled subjects. Missing at random assumption now entails that $f(Y|Z)$ for the sampled subjects also holds for the nonsampled subjects.

Now suppose that missing values in Y for the nonsampled subjects (that is, for all the $N - n$ subjects) are multiply imputed. Thus, each completed data is a "potential census" and there is no within-imputation variance and the combining rule then reduces to

$$\bar{Q}_{MI} = \sum_{l}^{M} Q_l/M,$$

$$T_{MI} = (1 + 1/M)B_M$$

where $B_M = \sum_{l}^{M}(Q_l - \bar{Q}_{MI})^2/(M-1)$ and the degrees of freedom for the reference t distribution is $\nu = (M-1)$. The fraction of missing information is 100% as a small portion of (sample) is used to fill a very large portion (nonsampled values).

From a practical point of view, one can set N to a large value compared to n such as $N = n/f$ where $f = 0.01$ or 0.001.

As an example consider data from an agricultural survey described in Sukhatme and Sukhatme (1970). The population for this survey consists of 170 villages in Lucknow subdivision in India. Cultivated area in 1931 (Z) is known for each of the 170 villages. A sample of 34 villages were selected in 1937 with probability proportional to size (cultivated areas in acres in 1931) and collected information about areas under wheat production (Y). The goal is to estimate the total cultivated area under wheat.

Figure 8.2 provides a scatter plot of $\sqrt{Y}$ and $\sqrt{Z}$ for 34 sample villages. Unfortunately, cultivated areas for the remaining (nonsampled) villages is not known except that the total cultivated area in the population is 78,019 acres and the sample total for 34 villages is 26,022 acres. Thus, the total cultivated area for the nonsampled population is (78,019-26,022)=51,997 acres.

Based on the scatter plot, and regression diagnostics, it is reasonable to fit a regression model,

$$\sqrt{Y} = \beta_o + \beta_1 \sqrt{Z} + \epsilon$$

with $\epsilon \sim N(0, \sigma^2)$. A normal distribution with mean 26.219 and standard deviation 8.654 seems to be a good fit for the marginal distribution of $\sqrt{Z}$. Assume a noninformative prior, $Pr(\beta_o, \beta_1, \sigma) \propto \sigma^{-1}$. The imputation step is as follows:

1. Sample $N - n = 170 - 34 = 126$ values of $\sqrt{Z}$ from a normal distribution with mean 26.319 and standard deviation 8.654. Multiply each value by a constant so that the total of the drawn values of Z for the 126 nonsampled villages equals 51,997. This may be viewed as raking to the known population total.

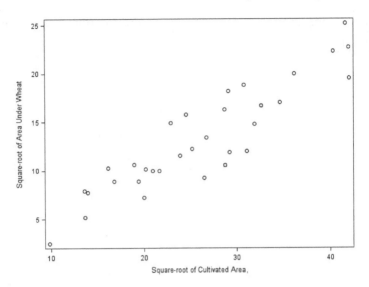

Figure 8.2: Scatter plot total cultivated area in 1931 and area under wheat in 1936, both on the square root scale

2. Draw values of σ, β_o and β_1 from their posterior distribution based on the regression of $\sqrt{Y}$ on $\sqrt{Z}$ using the sample data.

3. Draw from the predictive distribution of $\sqrt{Y}$ for the nonsampled subjects, given the draws in steps 1 and 2.

4. Compute the estimand Q.

5. Repeat steps 1-3, several times and apply the combining rules to obtain inference about Q.

The number of imputations should rather be large as the whole nonsampled population is being imputed. For this example analysis, 200 imputations yield the estimate of Q as 20,104, its standard error as 1,222 and the 95% interval estimates based on the t-distribution with 199 degrees of freedom as (17,709, 22,499). Sukhatme and Sukhatme (1970) provide a ratio estimate of 19,943 acres with the standard error of 1,242 acres.

8.4 Causal Inference

All causal inference problems may be viewed as a missing data problem when set up in the following potential outcome framework. Suppose that T is an indicator variable with 1 if the subject receives the treatment and 0 for placebo or control condition. At the outset of the experiment, there are two potential outcomes for each subject in the population consisting of N subjects. Let $Y_i(1)$ be the outcome that would be observed if the subject i were to receive the treatment and $Y_i(0)$ be the outcome that would be observed if the same subject were to receive the control condition. Since the subject i can receive either the treatment or control condition, actually only one of the two potential outcomes can be observed.

The causal effect for subject i, is the comparison of $Y_i(1)$ and $Y_i(0)$ (for example, the difference $U_i = Y_i(1) - Y_i(0)$ which is impossible to compute given that, only one of the two potential outcomes can be observed. The causal effect for the population, however, may be represented by the population mean $Q = \sum_i^N U_i/N = \sum_i Y_i(1)/N - \sum_i Y_i(0)/N = \bar{Y}(1) - \bar{Y}(0)$, the difference in the population means that would occur if all were given treatment versus if all were given the control condition. If the population size N were infinite, the estimand of interest is $\theta = E(Y(1) - Y(0))$. Sometimes, the causal effect for a subgroup based on $X = x$ may be of interest, $\theta_x = E[(Y(1) - Y(0))|X = x]$ where X is baseline covariates measured before assigning treatments.

Since the entire population cannot be observed, one or the other potential outcome can be observed only on a sample of subjects. Suppose that a representative sample of size $n + m$ is chosen at random, n subjects are given the treatment resulting in $Y_1(1), Y_2(1), \ldots, Y_n(1)$ and m subjects are given control condition resulting in $Y_{n+1}(0), Y_{n+2}(0), \ldots, Y_{n+m}(0)$. The resulting data structure is shown in Figure 8.3. This is the "file matching" data structure in Figure 1.1(c).

Suppose that the missing values in $Y(0)$ and $Y(1)$ are imputed separately. The marginal distributions of the observed and imputed values represent plausible distribution for the entire sample under the treated or control conditions, and thus allowing comparison of $f(Y(0))$ and $f(Y(1))$ accounting for any differences in X between the two groups.

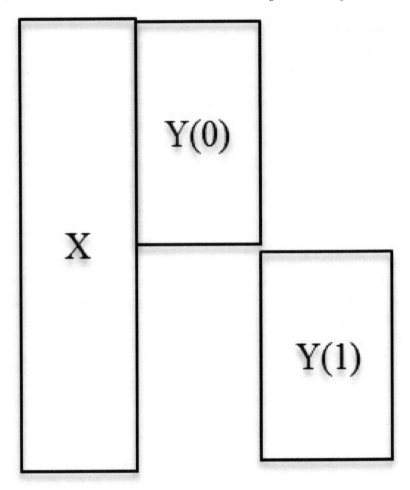

Figure 8.3: Data structure in the causal inference

The difference $U_i = Y_i(1) - Y_i(0)$ is not the only way to represent the individual level causal effect for subject i. Suppose one decides to use $U_i^* = (Y_i(1) - Y_i(0))/Y_i(0)$ to represent the individual level causal effect. Any aggregation of such measures across for population requires the joint distribution of $(Y(1), Y(0))$ or the conditional distribution of $(Y(1), Y(0))$ given X.

Obviously, there is no information in the data about the partial correlation coefficient between $Y(1)$ and $Y(0)$ conditional on X. The setup is very similar to the measurement error problem scenario (b) where W plays the role of X, Y plays the role of $Y(1)$ and X plays the role of $Y(0)$. However, the model assumptions made in that context may not be reasonable (that is, the assumption that $Pr(Y(1)|Y(0), X) = Pr(Y(1)|Y(0))$ may not be reasonable).

For concreteness, suppose that $Y(1)|X \sim N(\beta_o + \beta_1 X, \sigma_1^2)$ and $Y(0)|X \sim N(\alpha_o + \alpha_1 X, \sigma_o^2)$. Suppose that the correlation coefficient between $Y(1)$ and $Y(0)$ conditional on X is ρ and is known or fixed at some value. Assume a flat prior

$$Pr(\alpha_o, \alpha_1, \beta_o, \beta_1, \sigma_1 \sigma_o) \propto \sigma_1^{-1} \sigma_o^{-1},$$

for the remaining unknown parameters.

The following imputation scheme may be used to multiply impute the missing values in $Y(1)$ (for controls) and $Y(0)$ (for the treatment group).

1. Draw values of β_o, β_1 and σ_1 through a regression of $Y(1)$ on X.

2. Draw values of α_o, α_1 and σ_o through a regression of $Y(0)$ on X.

3. Compute $\mu_{1|0}(X) = \beta_o + \beta_1 X + \rho\sigma_1(Y(0) - \alpha_o - \alpha_1 X)/\sigma_o$, and $\sigma_{1|0}^2 = \sigma_1^2(1 - \rho^2)$.

4. Impute the missing values in $Y(1)$ by drawing from $N(\mu_{1|0}(X), \sigma_{1|0}^2)$.

5. Compute $\mu_{0|1}(X) = \alpha_o + \alpha_1 X + \rho\sigma_o(Y(1) - \beta_o - \beta_1 X)/\sigma_1$ and $\sigma_{0|1} = \sigma_o^2(1 - \rho^2)$.

6. Impute the missing values in $Y(0)$ by drawing from $N(\mu_{0|1}(X), \sigma_{0|1}^2)$.

Multiple imputation estimate of Q is the average of the completed data sample means of the paired differences U (or U^*) and the associated confidence intervals can be calculated using the standard combining rules. The confidence interval may be computed using the usual t reference distribution. By changing the values of ρ, sensitivity analysis can be displayed about the robustness of the causal effect.

8.5 Disclosure Limitation

Demand for access to data, especially collected using public funds, is ever growing. At the same time, availability of data from other sources such as web scraping, social media, administrative records and other public and private sources increases the chance of disclosure of a potential participant and their responses in a given study. This disclosure may be accidental or a deliberate attempt by an intruder who is snooping to find a particular subject. Nevertheless such disclosures violate the pledge of confidentiality often provided to the subjects during the informed consent process.

Multiple imputation framework can be used to alter the data before it is released but at the same time preserving some statistical content of the data. For example, suppose that the interest is in a regression analysis of a dependent variable Y on an independent variable X. Releasing the data may increase the chance of disclosure. It is possible to create (Y^*, X^*) (actually, a set of data sets) that preserves the distributional properties in the original data. That is, the inference about the population using (Y^*, X^*) will be "valid" in the same sense as the inference based on (Y, X). The term "valid" can be defined in terms of the repeated sampling properties where the estimates based on (Y^*, X^*) will be unbiased or consistent, and the interval estimates will have the stated coverage properties.

A heuristic definition of the term "valid" in the Bayesian sense is that the posterior distribution of a target population quantity based on (Y^*, X^*) is reflective of a plausible sample from the population. Thus, the results using (Y^*, X^*) is generalizable to the population just as (Y, X) is. There may be some loss of precision as (Y^*, X^*) will have more noise than the original data (Y, X) and hence a bit more uncertainty in the posterior distribution based on (Y^*, X^*) when compared the posterior distribution based on (Y, X). This also means that the sampling variance and confidence interval length may also be larger using (X^*, Y^*) when compared to (X, Y).

It is possible to generate (X^*, Y^*) such that some statistics $t(X^*, Y^*)$ equals $t(X, Y)$. For example, when (X, Y) is a bivariate normal, one can generate (X^*, Y^*) from a bivariate normal distribution such that the means, variances and the covariance exactly matches with those based on (X, Y). In many practical situations, it is difficult to match on statistics exactly. In general, the results based on (Y, X) will not be exactly the same as those based on (Y^*, X^*) but they may be "inferentially" equivalent.

The (Y^*, X^*) is called a set of fully synthetic data sets. Psychologically, it is hard to accept that one can get valid answers using synthetic data sets. Though, it is easily accepted that the two independent studies following the exact same design protocol will not produce the same answers but both are valid and "inferentially" equivalent. This psychological barrier is present even

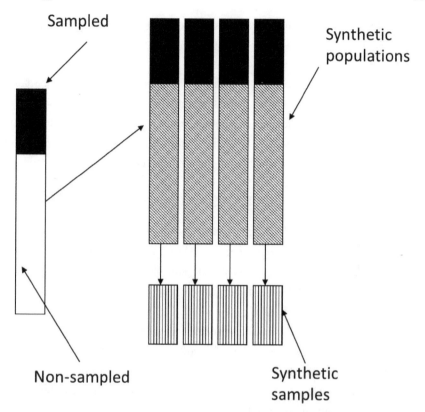

Figure 8.4: Schematic display of the process for creating fully synthetic data sets

among statisticians who routinely use bootstrap, Monte-Carlo methods for constructing inferences and are much comfortable with sampling variation.

There are obviously several limitations of this strategy. Not every feature in the actual data can be preserved in synthetic data sets. For example, if all the variables are categorical then some higher order interaction terms have to be set to zero while creating synthetic data sets in order to sufficiently perturb the actual values. That is, some statistical aspects, especially higher order interaction terms, will have to be sacrificed (for example, set to zero or a fixed value) while creating the synthetic data sets.

The principled justification of the synthetic data methodology, can be derived from the Bayesian perspective for inference from a finite population. Figure 8.4 provides a schematic display of generating synthetic data sets. Just like in a Bayesian inference section, the nonsampled values are treated as missing data and are multiply imputed. A subsample from each imputed data set is then obtained. This set of multiply imputed data sets form synthetic data sets.

The combining rules are different. Suppose $e_1, e_2, \ldots, e_M$ are the estimates from the M synthetic data sets and let $U_1, U_2, \ldots, U_M$ be the corresponding sampling variances. As before, let $\bar{e}_{MI} = \sum_l e_l/M$ be the mean and $B_M = \sum_l^M (e_l - \bar{e}_{MI})^2/(M-1)$ be the between-synthetic data variance of the estimates. The posterior variance is approximated by

$$T_{MI} = (1 + 1/M)B_M - \bar{U}_{MI}$$

where $\bar{U}_{MI} = \sum_l U_l/M$. This variance estimate can sometimes be negative, especially when the number of imputations, M, is small and the synthetic sample size is small. As the synthetic sample size approaches the population size, the synthetic data inference corresponds to the Bayesian inference discussed in Section 8.4 (as $\bar{U}_{MI} \to 0$).

The most straightforward approach for creating synthetic data sets using the standard multiple imputation software is to append k (the synthetic sample size) rows to the actual data set with all the variables set to missing. After the imputation drop all the actual records and retain only the k imputed records in the output data sets.

There are variations of this basic approach where not all variables are synthesized. Figures 8.5(a,b) illustrate two variations. Figure 8.5(a) corresponds to a situation where only some variables are synthesized. These may be sensitive variables or key variables that may be used in identification. The combining rules for the estimates are the same but the variance is given by $T_{MI} = \bar{U}_{MI} + B_M/M$.

An easy way to create these partially synthetic data sets is to append a copy of the original data to itself, set the key or sensitive variables to missing in the copy and then multiply impute the missing values in the appended data set. After imputation discard the original data set and treat the multiply imputed copies as partially synthetic data sets.

Figure 8.5(b) corresponds to a situation where a set of subjects who are deemed to be sensitive are mixed with nonsensitive subjects and only a selected set of variables for this mixed set are synthesized. The combining rules for the partial synthesis can be used in this situation. An easy way to create such a synthetic data set using the standard multiple imputation software is

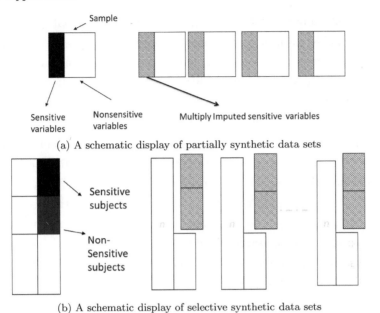

(a) A schematic display of partially synthetic data sets

(b) A schematic display of selective synthetic data sets

Figure 8.5: Variants of synthetic data sets to limit statistical disclosure

to append a copy of the original data to itself, set the key variables for the mixed set of subjects to missing, multiply impute and then retain only the multiply imputed copies as the synthetic data sets.

8.6 Bibliographic Note

Multiple imputation for measurement error problem is an obvious framework when combined with a validation sample. Raghunathan and Siscovick (1998) considers proxy and actual measures of exposure in a case-control analysis. Another early reference is Brownstone and Valletta (1996). Cole, Chu, and Greenland (2006) use this strategy in epidemiology. Padilla, Divers, Vaugh, Allison, and Tiwari (2009) consider a genetic application. Guo and Little (2011) consider multiple imputation approach for a regression analysis when covariates have heteroscedastic measurement error. Peytchev (2012) considers a survey context and Blackwell, Honacker, and King (2014) provides an overview of a unified approach for missing data and measurement error.

Multiple imputation for combining information from multiple data sources can be handled naturally using multiple imputation. Schenker, Raghunathan and Bondarenko (2010) and Raghunathan (2006) are examples using this approach for handling both measurement error and combining information. More systematic treatment in the survey context is considered in Dong (2012), Dong, Elliott, and Raghunathan (2014a, 2014b).

Ericson (1969) is probably the first systematic treatment of finite population inference from a Bayesian perspective. Basu (1971) also provides a discussion of the foundation of survey sampling inference. Chapter 2, in Rubin (1987) provides a comprehensive review of finite population inference. The book Little (2003) is useful to expand a view of finite population inference as a prediction of the nonsampled values.

Rubin (1974, 1978b) introduce the notion of potential outcomes in the causal inference. Related earlier work includes Neyman (1923) and Cox (1958). Holland (1986) is, perhaps, a starting place to understand the potential outcomes framework for causal inference and how it naturally leads to a missing data framework. Rubin (2005) also provides a comprehensive review of causal inference as it relates to missing data problem.

Generating fully synthetic data sets is proposed in Rubin (1993) and fully developed in Raghunathan, Reiter, and Rubin (2003). Partial synthesis is discussed in Little (1993) and fully developed in Reiter (2003). Also see Reiter (2005) for the full discussion of synthetic data sets for disclosure limitation. Little, Liu, and Raghunathan (2004) describes selective multiple imputation of key variables to avoid disclosure.

Finally, multiple uses of multiple imputation methodology is discussed in Reiter and Raghunathan (2007).

8.7   Problems

1. Fuller (1987) provides the following data on yields of corn (Y) on Marshall soil in Iowa for 11 sites along with the determination of available soil nitrogen, W. The soil nitrogen measurement are subject to error as it is based on a small sample from each site as well as the error in the chemical analysis. Suppose that it is known

that $X = W + \eta$ where $\eta \sim N(0, \tau^2 = 57)$. The goal is to fit the model $Y = \beta_o + \beta_1 X + \epsilon$ where $\epsilon \sim N(0, \sigma^2)$. Apply the multiple imputation methodology to construct point and interval estimates of β_1.

Site	Yield (Y)	Soil Nitrogen (W)	Site	Yield (Y)	Soil Nitrogen (W)
1	86	70	7	99	50
2	115	97	8	96	70
3	90	53	9	99	94
4	86	64	10	104	69
5	110	95	11	96	51
6	91	64			

Note that α_o, α_1 and τ^2 are known in the set of equations (8.2).

2. **Project:** Generate a sample of size 1000, $(X_i, Y_i, W_i), i = 1, 2, \ldots 1000$ from a trivariate normal distribution mean $(0, 1, 2)$ and covariance matrix,

$$\begin{bmatrix} 1 & 1 & 1.7 \\ & 2 & 1.5 \\ & & 4 \end{bmatrix}$$

(a) The goal is to infer about the regression coefficient for X in the model $Y = \beta_0 + \beta_1 X + \epsilon, \epsilon \sim N(0, \sigma^2)$. Set aside the first 900 observations as data from the main study and treat the remaining 100 observations as data from a substudy. Fit the above regression model on the main study data and store the point estimates of β_0, β_1 and σ and the interval estimate of β_1.

(b) Create a data corresponding to scenario (a) in Figure 8.1 by deleting X from the main study. Multiply impute the missing values of X in the main study. Perform multiply imputed analysis and again store the point and interval estimates of the same parameters.

(c) Create a data corresponding to scenario (b) by deleting X values from the main study and Y values from the substudy. Multiply impute the missing values of X in the main study. Perform multiply imputed analysis as in (b).

(d) Create a data corresponding to scenario (c) by deleting X on the main study, Y from the first 50 subjects in the substudy and W from the last 50 subjects. Perform multiply imputed analysis as in (b).

(e) Generate new samples and repeat the process (a) to (d), 250 times.

(f) Compare the bias and mean square properties of the estimates of β_o, β_1 and σ^2.

(g) Compute the true value of β_1 and calculate the actual coverage rate for each method of estimating the confidence interval. Also, calculate the length of the confidence intervals.

Based on this simulation study write a brief report on your findings and recommendations.

3. Refer to the example in Section 8.3.1. Suppose that one is interested in assessing the effect of not raking the generated values Z for the nonsampled subjects to the known population total. Construct inference about Q, by redoing the analysis of data without raking.

4. Sukhatme and Sukhatme (1970) provide area under wheat for 1936, X, as well. Construct inference about Q by using both X and Z as covariates.

5. **Project:** From the NHANES series, identify six years and a set of four variables (X_1, X_2, X_3, X_4) that were collected every year. Call this a complete data. From year 1, delete all values except for variables X_1 and X_2. Similarly, keep (X_1, X_3) in year 2; in year 3, (X_1, X_4)); in year 4 (X_2, X_3); in year 5, (X_2, X_4); and, finally, in year 6, keep (X_3, X_4). Vertically concatenate the six data sets. Multiply impute the missing values in the concatenated data set. Perform several example analysis that require 3 or more variables from the multiply imputed data sets and also on vertically concatenated six complete data sets. Compare the point and interval estimates. Write a brief report on your findings.

6. This problem is adapted based on a study described in Fleiss (1986). A large nursing home has a population of about 400 patients with senile dementia. Two methods (A and B) for training patients to take care of themselves were under consideration. A randomized study was conducted with 11 patients receiving training method A and 8 receiving training method B. Two weeks after training, each

patient was evaluated on 20 activities of daily living (unlocking a door, tying ones shoe laces, etc.) and the number of tests that were successful was obtained. The following table provides the number of successful tests on the nineteen patients in the study.

Method A	Method B
1, 3, 7, 5, 4, 1,	0, 3, 0, 1, 0, 0
2, 1, 6, 1, 5	1, 2

(a) Suppose that the training method A was imparted to all 400 patients then infer about the total number tasks the population of patients will be able to take care of themselves?

(b) Answer the same question as in (a) for the training method B.

(c) Infer about the increase in the number of tasks that a typical patient would be able to take care of himself/herself by providing the training method A over B.

(d) Infer about the proportionate increase in the number of tasks that a typical patient would be able to take care of himself/herself by providing the training method A over B.

Clearly state all the model assumptions. Write a brief report on advantages of the training method A over B.

7. **Project**. Consider any data set discussed in this book. Create a set of fully synthetic data sets and fit the same model on both the actual and synthetic data sets. Compare the results and then write a report on the utility of multiple imputation procedure for disclosure limitation.

9

Other Topics

9.1 Uncongeniality and Multiple Imputation

A typical multiple imputation framework involves two parties or individuals: An imputer and the analyst. An imputer working with a set of variables develops predictive models for the missing values conditional on the observed values and generates imputations. On the other hand, the analyst, perhaps uses a subset of variables or even a subset of subjects for the analysis using a model that may differ from the predictive model used by the imputer. It is possible that the analyst uses the imputer model but settles on a statistically less efficient procedure for convenience. These differences between the imputer and analyst introduce "uncongeniality." The goal is to assess how such uncongeniality affects the inferences for the analyst using multiply imputed data sets.

Let e_{MI} be the multiple imputation estimate of a parameter, θ, in the analyst model constructed based on the completed data sets created by an imputer. Let T_{MI} be the corresponding multiple imputation variance estimate. Let E_A and V_A denote the repeated sampling expectation and variance calculated under the analyst model. Ideally, one desires that $E_A(e_{MI}) = \theta$ or at the least consistency where $|e_{MI} - \theta|$ converges to zero as the sample size tends to infinity and the fraction of missing information converges to a fixed number, $\lambda < 1$. Also, one would desire an unbiased variance estimate, $V_A(e_{MI}) = E_A(T_{MI})$, or at least a consistent variance estimate where $|V_A(e_{MI}) - E_A(T_{MI})|$ converges to zero under the same conditions for the parameter estimate stated above.

This may not happen in some situations as discussed in Chapter 5. For example, suppose that the imputer used a regression model to impute the missing values in three variables Y, X and Z. However, the analyst is interested in the relationship between Y and Z and *assumes* that X is unrelated to Y and Z. That is, from the analyst perspective, the imputer has used a "noise" (X) in the imputation process. The multiple imputation estimate, though, consistent is likely to be inefficient and the variance estimate to be conservative in that $V_A(e_{MI}) \leq E_A(T_{MI})$. One the other hand, if the imputer model were to be a better reflection of reality then the multiple imputation estimate can be more desirable or efficient.

On the other hand, if the imputer used only (Y, Z) in the imputation process and the analyst is interested in the analysis involving (Y, X, Z) and the repeated sampling calculations were to be conducted under the analyst model then the estimates are biased as imputer has assumed no relationship between (Y, Z) and X. The BLS simulation study in Chapter 5 is an example of this situation.

A more complex situation of uncongeniality can be illustrated using the following example. Consider a bivariate normal random sample of size n where X is fully observed, and Y is observed on the first m subjects and is missing for the remaining $n - m$ subjects. The goal is to estimate $\theta = Pr(Y \leq 1)$. The imputer generates imputations by drawing values from the predictive distribution based on a regression model $Y = \beta_o + \beta_1 X + \epsilon$ where $\epsilon|X \sim N(0, \sigma^2)$ and a flat prior $\pi(\beta_o, \beta_1, \sigma) \propto \sigma^{-1}$.

The analyst assumes the same bivariate normal model but estimates θ using $\bar{Z} = \sum_i Z_i/n$ where $Z_i = 1$ if $Y \leq 1$ and 0 otherwise. Let $\bar{Z}_l$ be the estimate based on completed data l and $U_l = \bar{Z}_l(1 - \bar{Z}_l)/n$ be the corresponding completed data variance estimate. As usual, let $\bar{Z}_{MI}$ and $T_{B,MI}$ be the multiple imputation estimate and its variance estimate where B subscript indicates the binary version of Y being used in constructing the inference.

On the other hand, the asymptotically efficient estimate of θ from the completed data l is $\widehat{\theta}_l = \Phi([1 - \bar{Y}_l]/s_{Yl})$ where $\bar{Y}_l$ and s_{Yl} are the mean and the standard deviation, respectively, and $\Phi()$ is the cumulative distribution function of the standard normal distribution. Using Taylor's series expansion,

$$\widehat{\theta}_l = \theta - (\bar{Y}_l - \mu_Y)\phi([1 - \mu_Y]/\sigma_Y)/\sigma_Y - (s_{Yl}^2 - \sigma_Y^2)\phi([1 - \mu_Y]/\sigma_Y)/(2\sigma_Y^3),$$

where $\phi(.)$ is the density function of the standard normal distribution. The variance of the estimate can be shown to be

$$U_l = a_l^2[1/n + 1/(2s_{Yl}^2(n - 1))],$$

where $a_l = \phi[(1 - \bar{Y}_l)/s_{Yl}]$. Let $\bar{\theta}_{MI}$ and $T_{E,MI}$ be the asymptotically efficient estimate and its multiple imputation variance estimate. The asymptotically efficient estimate of θ satisfies the assumptions given in Section 4.5 whereas the analyst estimation procedure does not.

It can be shown that the expected value of $T_{B,MI}$ under repeated sampling from a bivariate normal distribution tends to be larger than $var(\bar{Z}_{MI})$ whereas the expected value of $T_{E,MI}$ is nearly equal to $var(\hat{\theta}_{MI})$. That is, conservative inferences are obtained using the multiple imputation procedure if the analyst were to use a less than efficient estimation procedure on completed data sets.

A question arises: "What is the correct sampling distribution under which inference based on $(\bar{Z}_{MI}, T_{B,MI})$ should be evaluated"? From the analyst perspective, the calculation should perhaps be made based on repeated sampling of (X, Z) where Z is imputed using a probit model.

Now consider a more serious uncongeniality problem. Suppose that the analyst model is $Y = \beta_o + \beta_1 X + \epsilon$ where $\epsilon|X \sim N(0, \sigma^2 X^\alpha)$ and this model fits the data well. However, the imputer ignored the heteroscedasticity in the error distribution. This type of uncongeniality can introduce bias in the estimate of θ, and it is not that meaningful to assess the variance properties of biased estimates. In general, "under fitting" by imputer relative to the analyst model (assuming that the repeated sampling calculations are performed under the analyst model) will lead to biased inferences. Thus, the general advice for any imputer is to use as many variables as possible and incorporate structures in the observed data in the imputation process and to make sure that imputation models are a reasonably good fit for the data.

Uncongeniality is of less concern as long as e_{MI} is a consistent estimator of θ and $E_A(T_{MI}) \geq V_A(e_{MI})$. If the imputations were obtained using good fitting predictive models then uncongeniality may not be a serious issue. On the other hand, if the imputer ignores some important features while creating predictive models then multiple imputation inference can be biased.

9.2   Multiple Imputation for Complex Surveys

Much of applied research is conducted using data collected from a probability sample of subjects. Often these surveys employ unequal probabilities of selection based on some covariates leading to survey weights, clustering due to administrative convenience, or even a necessity, of conducting survey interviews and stratification to ensure that all sections of the society are represented in the sample. These features are usually related to the analytical variables collected from respondents and, therefore, should be included in the imputation process.

The survey weights may be included as predictors in both modeling the mean and the variance function of the variables being imputed. To maintain the stratification, separate imputations may be performed within each stratum. The case-control study is an example with two strata (cases and controls). Alternatively, dummy variables for stratum could be used as predictors.

More complex modeling is involved with clustering. If the observations within a cluster are highly correlated then the imputations should also reflect the same degree of correlation. A random effects model may be used to incorporate such correlations. There are no easily available software packages for performing imputations under such models.

There are two main aspects of clustering that need to be distinguished. The first aspect is the correlation induced by the survey design and is more a function of design based calculations, such as sampling variances. The second aspect is the substantive nature of correlation among the variables. To illustrate these two aspects, consider the two design options. In the first design, block groups are sampled, and then households are listed, sampled and one person from each sampled household is selected for the interview. In the second design, the same strategy is used except that half the number of households are sampled and two persons are sampled from each household. The correlation among observations is a more serious issue in the second design option than in the first. For the second design, one may view the data as repeated measures at the household level for imputing the missing values.

In practice, using the weight variable, stratum and clusters as dummy variables with possible interaction terms with some survey variables as predictors in the imputation model should be sufficient. More elaborate models using random effects may be needed to incorporate strong correlation among observations from a substantive point of view.

9.3 Missing Values by Design

Many large scale studies have competing goals and may result in a long questionnaire. Such a long questionnaire may result in break-offs, item nonresponse and poor quality data due to respondent fatigue. It is not uncommon to find that after a few items, with say five response options, have been administered, the respondents begin to choose middle response options more frequently. One design option is to be selective or tailor the questions asked from each respondent.

Matrix or multiple matrix sampling designs, used in market research, randomly select items to be administered for each sampled subject. A disadvantage of completely random selection is that some parameters may not be estimable. For instance, if two items are never administered on the same set of subjects then it is not possible to estimate the correlation between them. Thus, over the past several years, the random selection of items to be administered has been replaced by designed allocation of items to subjects so that key parameters or target quantities can be estimated from the resulting data. Implementation of such designs also has become easier with the advent of computer assisted interviewing process.

Suppose that a questionnaire consists of p items and is split into a core component (core split) with m items and the remaining $p - m$ items split into k components (splits). A split questionnaire survey design involves administering the core split and a subset from the k splits. For example, a sampled subject is administered the core split plus two out of k splits in such a way that every $k(k-1)/2$ pairs are administered to the same number of subjects. Suppose $k = 4$ and the sample size is $n = 600$. The questionnaire configuration $(C, 1, 2)$ is assigned to 100 subjects at random, $(C, 1, 3)$ to another 100, ..., and $(C, 3, 4)$ to the last remaining 100 subjects. This will allow estimation of means, variances and correlation coefficients between any two of the p items, and hence, the regression coefficients. The length of the questionnaire can be reduced by having more splits and still administering only two splits but for a fewer number of subjects. For example, $k = 5$ would result in 10 questionnaire configurations each randomly assigned to 60 sampled subjects.

Note that, by design, the data are missing completely at random. Missing values in the resulting data set can be multiply imputed and analyzed using methods described in Chapters 3 and 4. Creation of split questionnaires requires some thought. Some items have to stay together to maintain contextual consistency. Generally, the items included in a particular split should be (highly?) predictive of items included in other splits. By assigning items to splits such that the within-split correlation is low and the between-split correlation is high may lead to a low fraction of missing information. Having a pilot or preliminary data can help in developing a proper split questionnaire. Another option is an adaptive approach where the full questionnaire is implemented initially, use the data collected to develop splits and then switch over to the split questionnaire design.

9.4   Replication Method for Variance Estimation

Previous chapters described bootstrap and jackknife approaches (two approaches from a broad set called replication techniques) for variance estimation under a variety of contexts. It is possible to use these two approaches, especially in the context of complex surveys involving clustering, to compute the variance estimate based on a single imputation (as an alternative to the multiple imputation method). Basically, the three steps are: (1) generate a replicate, (2) singly impute the missing values using an appropriate method, and (3) compute the point estimate of the parameter of interest. Repeat the process several times and use the replicate point estimates to compute the variance estimate.

The replication procedure will be described for imputing missing values in the data from a two-stage cluster sample design. The population is divided into C clusters with N_j units in cluster $j = 1, 2, \ldots, C$. A sample of c clusters is obtained from C clusters with probability proportional to size, $N_j, j = 1, 2, \ldots, C$. Without any loss of generality, assume that the sampled clusters are labeled $j = 1, 2, \ldots, c$ and the nonsampled clusters are $c+1, c+2, \ldots, C$. From each sampled cluster, a simple random sample of size $n_j = f \times N_j, j = 1, 2, \ldots, c$ is obtained where f is a fixed number less than 1. Note that, every subject in the population has the same probability of being included in the sample. Such designs are called Equal Probability SElection Method (EPSEM) designs. Assume that N_j is relatively large compared to n_j.

Suppose that the two variable of interest are (Y, X) which are approximately bivariate normal. Inferential quantities of interest are the population mean and standard deviation of X and the regression coefficient in the model $Y = \alpha_o + \alpha_1 X + \epsilon$. In the survey, suppose that every subject provides Y but only $r_j, j = 1, 2, \ldots, c$ subjects in cluster j provide information on X. Assume that the data are missing at random in that $Pr(X \text{ is Missing}|Y) = g(Y)$ and is the same across all the clusters.

The analyst chooses a procedure for a single imputation of the missing values in X. Once the missing values have been imputed, the analyst plans to apply standard complete-data methods to estimate the target quantities of interest. A replication procedure for incorporating imputation uncertainty involves the following steps:

1. Create a replicate by deleting a cluster.

2. Apply the imputation procedure to impute the missing values in the replicate.

3. Apply the complete data method on the completed-replicate data set to obtain the point estimates of the parameters of interest.

Let q_l be the estimate of the population quantity, Q, from replicate $l = 1, 2, \ldots, R$. In this example, $R = c$, the number of sampled clusters. The sampling variance is computed as

$$SE_R^2 = (R - 1) \sum_l (q_l - \bar{q})^2 / R,$$

where $\bar{q} = \sum_l q_l / R$ is the mean across the R replicates. Note that the imputation changes across the replicates and, hence, captures the imputation uncertainty. The replication procedure is appealing, if the imputation task is not too onerous to apply on each replicate or if it is difficult to obtain the completed-data sampling variance, required in the multiple imputation approach. This approach, however, is computationally more intensive than the multiple imputation approach.

It is important to select the imputation procedure that will yield valid estimates of all the population quantities of interest. In this example, the imputation procedure should account for clustering, yield valid inferences for the population mean, standard deviation and the regression analysis of Y on X. The methods described in Chapter 3 can be used to perform imputation within each cluster provided the sample size is large. When the cluster sample sizes are small, regression models with random cluster effects may be used to preserve the correlation between observations within the same cluster in the imputation process. The estimates of the population quantities may be biased, if the imputation procedure is not carefully chosen.

Suppose that the analyst uses the mean of all the respondents in the survey as the imputed value for each nonrespondent. This can result in a biased estimate of the parameters and, hence, rendering the replication standard error as a "valid standard error of the invalid estimate". Thus, regardless which procedure, replication or multiple imputation, is used to incorporate the imputation uncertainty, the imputation method must be carefully chosen to yield a proper or a valid estimate of the population quantities of interest.

9.5   Final Thoughts

Missing data introduces complexity in statistical inference about an unknown parameter or the population quantity. The complete data model assumption and missing data mechanism are needed to construct inference from incomplete data. The general goal is to infer about the parameter, by using the observed information on the variables of interest, incorporating information from other variables about the missing values and using reasonable model assumptions. This can be supplemented by performing sensitivity analysis with respect to the assumptions. One has to be clear by answering three fundamental questions in the order: "Do you have missing values and, if so, what to do with them.?" Once you have the answer then ask "Why does this answer makes sense"? Finally, proceed towards answering "How to implement the plan firmed by answering the two previous questions"?

There are no assumption free approaches. Weighting, imputation and observed data-likelihood are three approaches making different assumptions. Multiple imputation approach serves as the most general purpose method. Is the multiple imputation the absolute best methodology? Of course, no. In statistical inference, where one revels in uncertainties, having an absolute point of view is untenable. It is by far the most practical method that can be easily implemented because of readily available software. However, care needs to be taken in developing empirically tested prediction models using available or adapting model diagnostic tools.

Several articles have appeared demonstrating multiple imputation variance estimator can be invalid under some conditions. This may be due to the analyst not using fully efficient methods or due to imputer ignoring variables of interest to the analyst. Some of these situations were considered in Section 9.1. Generally, the multiple imputation variance estimators are conservative, leading to over coverage (that is, 95% confidence intervals, for example, may actually include the true value more than 95% of the time). There are situations where the converse can be true. From a practical point of view, a carefully chosen imputation process and a decently efficient statistical method for the analysis of completed data will not have such problems. Hope that material presented in the book is useful and good luck with your next analysis of data with missing values!!

9.6 Bibliographic Note

Fay (1992) was the first to demonstrate potential bias in the multiple imputation variance estimator when the model used by the imputer is not congenial with the estimate used by an analyst, though the term congenial and uncongenial in the multiple imputation context was introduced by Meng (1994). Fay (1996), Rao (1996) and Rubin (1996) and discussion by several authors consider several issues and alternative methods when the analyst and imputer use different models or procedures, sources of inputs, etc. Wang and Robins (1998) and Robins and Wang (2000) through several examples explore the nature of multiple imputation variance estimator and suggest their own version of variance estimator. This variance estimator is much harder to compute and requires information that is not usually produced by software packages. Kim et al (2006) discusses the problem with Rubin's variance estimator with uncongeniality and suggest some alternatives such as fractional imputation. Rubin (2003) responds to these criticisms.

Matrix or multiple matrix sampling designs is discussed in Shoemaker (1973). This type of design has been used in educational assessment surveys (see, for example, Mislevy et al (1992), Thomas and Gan (1997), Zeger and

Thomas (1997). Raghunathan and Grizzle (1995) developed the split questionnaire design in the health survey context and used multiple imputation for the analysis of data from such design. Thomas et al (2006) discuss methods for constructing split questionnaire designs. Wacholder et al (1994) use a similar idea to create partial questionnaire designs. See also Little and Rhemtulla (2013) for a review.

Rubin (1987) discusses the importance of including the design variables such as survey weights, clustering and stratification in the imputation model, especially, if they are predictive of survey outcome with missing values. Reiter, Raghunathan and Kinney (2006) demonstrate the importance of including the design variables in the imputation model and some strategies for doing so. Sometimes it is difficult to fit models for commonly used designs such as selection of two primary sampling units (clusters) within a stratum and where the sample sizes within the sampling units may be relatively small. More promising and easily implementable methods are given in Zhou (2014). A systematic development of jackknife repeated replication method for computing variance estimates with singly imputed data is given in Rao and Shao (1992). Shao (1996) provide provides a review of resampling methods in sample surveys. Another good source to learn more about these methods is Kim and Shao (2014).

9.7   Exercises

1. **Project:** Generate 200 samples each of size 250 under the following model assumptions: $X \sim Ber(1, 0.5)$ and $Y|X \sim N(1, 1)$. Set some values of Y to missing using an MCAR mechanism where the missing data indicator $R \sim Ber(1, 0.4)$. Multiply impute ($M = 10$ imputations) the missing values in Y using the model $Y \sim N(\mu, \sigma^2)$ with the prior $Pr(\mu, \sigma) \propto \sigma^{-1}$.

   (a) The analyst, unaware of the true model, would like to estimate the population mean for the subset defined by $X = 1$. Construct before-deletion, complete-case and multiply imputed point estimates of the subpopulation mean. Compute the bias and mean-square errors of these three point estimates.

(b) Compute before deletion, complete case and multiple imputation confidence interval and its length. Compare the average length and coverage properties.

(c) Summarize your findings and comment on the effect of uncongeniality.

2. **Project:** For the same data sets generated in problem 1, use the following imputation model $Y|X \sim N(\beta_o + \beta_1 X, \sigma^2)$ and a prior $Pr(\beta_o, \beta_1, \sigma) \propto \sigma^{-1}$. Repeat items (a)-(c) as in problem 1. Discuss the recommendation "An imputer should condition on all the variables (even the irrelevant) that could be used by the Analysts."

3. Several homework problems involved downloading NHANES data and performing multiple imputation analysis. Redo those analyses by incorporating weights, clustering and stratification using some of the strategies discussed in Section 9.2. Compare the results with and without incorporating the design variables as predictors.

Bibliography

[1] Afifi, A. A., & Elashoff, R. M. (1966). Missing observations in multi-variate statistics: I. Review of the literature. *Journal of the American Statistical Association*, 61, 595-604.

[2] Afifi, A. A., & Elashoff, R. M. (1967). Missing observations in multivariate statistics: II. Point estimation in simple linear regression. *Journal of the American Statistical Association*, 62, 10-29.

[3] Allan, F. E., & Wishart, J. (1930). A method of estimating the yield of a missing plot in field experimental work. *Journal of Agricultural Science*, 20, 399-406.

[4] Amemiya, T. (1984). Tobit models: A survey. *Journal of Econometrics*, 24, 3-61.

[5] Anderson, T. W. (1957). Maximum likelihood estimates for a multivariate normal distribution when some observations are missing. *Journal of the American Statistical Association*, 52, 200-203.

[6] Andridge, R. R., & Little, R. J. A. (2010). A review of hot deck imputation for survey non-response. *International Statistical Review = Revue Internationale De Statistique*, 78, 40-64.

[7] Baker, S. G., & Laird, N. M. (1988). Regression analysis for categorical variables with outcome subject to nonignorable nonresponse. *Journal of the American Statistical Association*, 83, 62-69.

[8] Barnard, J., & Rubin, D. (1999). Miscellanea. Small-sample degrees of freedom with multiple imputation. *Biometrika*, 86, 948-955.

[9] Basu, D. (1971). An essay on the logical foundations of survey sampling, part I (with discussion). In V. P. Godambe, & D. A. Sprott (Eds.), *Foundations of Statistical Inference (pp. 203-243)*. Toronto: Holt, Rinehart and Winston.

[10] Bishop, Y. M., Fienberg, S. E., & Holland, P. W. (2007). *Discrete Multivariate Analysis: Theory and Practice*. Springer Science & Business Media.

[11] Blackwell, M., Honaker, J. A., & King, G. (2015). A unified approach to measurement error and missing data: Overview and Applications, *Sociological Methods and Research, Sociological Methods and Research*, 1-39.

[12] Bose, S. S. (1938). Appendix. the estimation of mixed-up yields and their standard errors. *Sankhy: The Indian Journal of Statistics (1933-1960)*, *4*, 112-120.

[13] Bose, S. S., & Mahalanobis, P. C. (1938). On estimating individual yields in the case of mixed-up yields of two or more plots in field experiment. *Sankhy: The Indian Journal of Statistics (1933-1960)*, *4*, 103-111.

[14] Box, G. E. P., & Tiao, G. C. (1973). *Bayesian inference in statistical analysis*. New York: Wiley Classics.

[15] Brand, J. P. L. (1999). *Development, implementation and evaluation of multiple imputation strategies for the statistical analysis of incomplete data sets*. (Ph.D. Thesis, Erasmus University, Rotterdam).

[16] Brick, M. J. (2013). Unit nonresponse and weighting adjustments: A critical review. *Journal of Official Statistics, 29*, 329-353.

[17] Brownstone, D., & Valletta, R. G. (1996). Modeling earnings measurement error: A multiple imputation approach. *The Review of Economics and Statistics, 78*, 705-717.

[18] Cao, W., Tsiatis, A. A., & Davidian, M. (2009). Improving efficiency and robustness of the doubly robust estimator for a population mean with incomplete data. *Biometrika, 96*, 723-734.

[19] Carpenter, J., & Kenward M. (2013). *Multiple Imputation and its Application*, West Sussex, UK, John Wiley and Sons.

[20] Casella, G., & Berger, R. L. (2002). In Crockett C. (Ed.), *Statistical Inference* (2nd ed.). Pacific Groves, CA, United States: Duxbury Press.

[21] Chambers, R. L., & Skinner, C. J. (2003). *Analysis of Survey Data*. New York: John Wiley and Sons.

[22] Cole, S. R., Chu, H., & Greenland, S. (2006). Multiple-imputation for measurement-error correction. *International Journal of Epidemiology, 35*, 1074-1081.

[23] Cox, D. R. (1958). Some problems connected with statistical inference. *The Annals of Mathematical Statistics, 29*, 357-372.

[24] Cox, D. R., & Hinkley, D. V. (1979). *Theoretical Statistics*. Nachtsheim, C., Boca Raton, FL CRC Press.

[25] Davidian, M., Tsiatis, A. A., & Leon, S. (2005). Semiparametric estimation of treatment effect in a Pretest–Posttest study with missing data. *Statistical Science, 20*, 24 August 2005-261-301.

[26] Deming, W. E. (1950). *Some Theory of Sampling*. New York: John Wiley & Sons.

[27] Dempster, A. P., Laird, N. M., & Rubin, D. B. (1977). Maximum likelihood from incomplete data via the EM algorithm. *Journal of the Royal Statistical Society. Series B (Methodological), 39*, 1-38.

[28] Dempster, A. P., Rubin, D. B., & Tsutakawa, R. K. (1981). Estimation in covariance components models. *Journal of the American Statistical Association, 76*, 341-353.

[29] Deville, J. C., Srndal, C. E., & Sautory, O. (1993). Generalized raking procedures in survey sampling. *Journal of the American Statistical Association, 88*, 1013-1020.

[30] Diggle, P., Heagerty, P., Liang, K., & Zeger, S. (2002). *Analysis of Longitudinal Data*. Oxford, UK, Oxford University Press.

[31] Dodge, Y. (1985). *Analysis of Experiments with Missing data*. New York: John Wiley & Sons.

[32] Dong, Q. (2012). Combining information from multiple complex surveys. (Doctoral dissertation, University of Michigan, 2012).

[33] Dong, Q., Elliott, M. R., & Raghunathan, T. E. (2014a). A nonparametric method to generate synthetic populations to adjust for complex sampling design features. *Survey Methodology, 40*, 29-46.

[34] Dong, Q., Elliott, M. R., & Raghunathan, T. E. (2014b). Combining information from multiple complex surveys. *Survey Methodology, 40*, 347-354.

[35] Draper, N. R., & Smith, H. (1998). *Applied regression analysis* (3rd ed.). New York: John Wiley and Sons.

[36] Efron, B. (1994). Missing data, imputation, and the bootstrap. *Journal of the American Statistical Association, 89*, 463-475.

[37] Elderton, E. M., & Pearson, K. (1910). *A first study of the influence of parental alcoholism on the physique and ability of the offspring* (2nd ed.). London: Eugenics Laboratory Memoirs X, University of London, Francis Galton Laboratory for National Eugenics. Dulau and Co. Limited.

[38] Elliott, M. (2007). Bayesian weight trimming for generalized linear regression models. *Survey Methodology, 33*, 23.

[39] Elliott, M. R. (2008). Model averaging methods for weight trimming. *Journal of Official Statistics, 24*, 517-540.

[40] Elliott, M. R. (2009). Model averaging methods for weight trimming in generalized linear regression models. *Journal of Official Statistics, 25*, 1-20.

[41] Elliott, M. R., & Little, R. J. A. (2000). A Bayesian approach to combining information from a census, a coverage measurement survey, and demographic analysis. *Journal of the American Statistical Association, 95*, 351-362.

[42] Ericson, W. A. (1969). Subjective Bayesian models in sampling finite populations. *Journal of the Royal Statistical Society. Series B (Methodological), 31*, 195-233.

[43] Fay, R. E. (1992). When are inferences from multiple imputation valid? *Proceedings of the Survey Research Methods Section of the American Statistical Association,* pp. 227-232.

[44] Fay, R. E. (1996). Alternative paradigms for the analysis of imputed survey data. *Journal of the American Statistical Association, 91*, 490-498.

[45] Fleiss, J. L. (1986). *The Design and Analysis of Clinical Experiments.* New York: John Wiley & Sons.

[46] Fleming, T. R., & Harrington, D. P. (2005). *Counting Processes and Survival Analysis.* Hboken, New Jersey: John Wiley & Sons.

[47] Ford, B. (1983). An overview of hot-deck procedures. In W. Madow, I. Olkin & D. Rubin (Eds.), *Incomplete data in sample surveys* (2nd ed., pp. 185-207). New York: Academic Press.

[48] Fuller, W. A. (1987). *Measurement Error Models.* New York: John Wiley & Sons.

[49] Gelman, A., & Hill, J. (2006). *Data Analysis using Regression and Multilevel/Hierarchical Models.* New York: Cambridge University Press.

[50] Gelman, A., Carlin, J. B., Stern, H. S., Dunson, D. B., Vehtari, A., & Rubin, D. B. (2014). *Bayesian Data Analysis, third edition.* Boca Raton, FL: CRC Press, Taylor & Francis Group.

[51] Glasser, M. (1964). Linear regression analysis with missing observations among the independent variables. *Journal of the American Statistical Association, 59*, 834-844.

[52] Glasser, M. (1965). Regression analysis with dependent variable censored. *Biometrics, 21*, 300-307.

[53] Glynn, R., Laird, N., & Rubin, D. (1986). Selection modeling versus mixture modeling with nonignorable nonresponse. In H. Wainer (Ed.), (pp. 115-142) Springer, New York.

[54] Graubard, B. I., & Korn, E. L. (1999). Predictive margins with survey data. *Biometrics, 55*, 652-659.

[55] Guo, Y., & Little, R. J. (2011). Regression analysis with covariates that have heteroscedastic measurement error. *Statistics in Medicine, 30*, 2278-2294.

[56] Hartley, H. O. (1946). Discussion on "A review of recent statistical developments in sampling and sampling surveys" by F. Yates. *Journal of the Royal Statistical Society: Series A (Statistics in Society), 109*, 37-38.

[57] Hartley, H. O. (1958). Maximum likelihood estimation from incomplete data. *Biometrics, 14*, 174-194.

[58] Hartley, H. O., & Hocking, R. R. (1971). The analysis of incomplete data. *Biometrics, 27*, 783-823.

[59] Heckman, J. J. (1976). The common structure of statistical models of truncation, sample selection and limited dependent variables and a simple estimator for such models. *Annals of Economic and Social Measurement* (Sanford V. Berg ed., pp. 475-4925) NBER.

[60] Heeringa, S. G., West, B. T., & Berglund, P. A. (2010). *Applied Survey Data Analysis*. Boca Raton, FL: Chapman & Hall/CRC.

[61] Heyting, A., Tolboom, J. T., & Essers, J. G. (1992). Statistical handling of drop-outs in longitudinal clinical trials. *Statistics in Medicine, 11*, 2043-2061.

[62] Hogg, R. V., KcKean, J., & Craig, A. T. (2012). *Introduction to Mathematical Statistics* (7th ed.). U.S.A.: Pearson Education Limited.

[63] Holland, P. W. (1986). Statistics and causal inference. *Journal of the American Statistical Association, 81*, 945-960.

[64] Holt, D., & Smith, T. M. F. (1979). Post stratification. *Journal of the Royal Statistical Society. Series A (General), 142*, 33-46.

[65] Horsley, V. D. (1911). *Alcohol and the Human Body an Introduction to the Study of the Subject, and a Contribution to National Health,* (Ed. 4. ed.). London: Macmillan and Co.

[66] House, J.,S. (2014). Americans' changing lives: Waves I, II, III, IV, and V, 1986, 1989, 1994, 2002, and 2011. *Ann Arbor, MI: Inter-University Consortium for Political and Social Research [Distributor], ICPSR04690-v7*

[67] Izrael, D., Hoaglin, D., & Battaglia, M. (2000). A SAS macro for balancing a weighted sample. *Proceedings of the Twenty-Fifth Annual SAS*

Users Group International Conference, Cary, NC: SAS Institute Inc.
pp. 1350-1355.

[68] Jones, M. P. (1996). Indicator and stratification methods for missing explanatory variables in multiple linear regression. *Journal of the American Statistical Association, 91*, 222-230.

[69] Kaciroti, N. A., & Raghunathan, T. (2014). Bayesian sensitivity analysis of incomplete data: Bridging pattern-mixture and selection models. *Statistics in Medicine, 33*, 4841-4857.

[70] Kaciroti, N. A., Raghunathan, T. E., Schork, M. A., & Clark, N. M. (2008). A Bayesian model for longitudinal count data with non-ignorable dropout. *Journal of the Royal Statistical Society: Series C (Applied Statistics), 57*, 521-534.

[71] Kaciroti, N. A., Raghunathan, T. E., Schork, M. A., Clark, N. M., & Gong, M. (2006). A Bayesian approach for clustered longitudinal ordinal outcome with nonignorable missing data: Evaluation of an asthma education program. *Journal of the American Statistical Association, 101*, 435-446.

[72] Kaciroti, N. A., Schork, M. A., Raghunathan, T., & Julius, S. (2009). A Bayesian sensitivity model for intention-to-treat analysis on binary outcomes with dropouts. *Statistics in Medicine, 28*, 572-85.

[73] Kalbfleisch, J. D., & Prentice, R. L. (2002). *The Statistical Analysis of Failure Time Data* (2nd ed.). New York: John Wiley and Sons.

[74] Kalton, G., & Flores-Cervantes, I. (2003). Weighting methods. *Journal of Official Statistics, 19*, 81-97.

[75] Kennickell, A. B. (1991). Imputation of the 1989 survey of consumer finances: Stochastic relaxation and multiple imputation. *Proceedings of the Survey Research Methods Section of the American Statistical Association*, pp. 1-10.

[76] Kermack, W. O., & McKendrick, A. G. (1991). Contributions to the mathematical theory of epidemics–I. 1927. *Bulletin of Mathematical Biology, 53*(1-2), 33-55.

[77] Kermack, W. O., & McKendrick, A. G. (1991). Contributions to the mathematical theory of epidemics–II. the problem of endemicity.1932. *Bulletin of Mathematical Biology, 53*(1-2), 57-87.

[78] Kermack, W. O., & McKendrick, A. G. (1991). Contributions to the mathematical theory of epidemics–III. further studies of the problem of endemicity. 1933. *Bulletin of Mathematical Biology, 53*(1-2), 89-118.

[79] Kim, J. K., & Shao, J. (2014). *Statistical Methods for Handling Incomplete Data.* Boca Raton: CRC Press, Taylor & Francis Group.

[80] Kim, J. K., Michael Brick, J., Fuller, W. A., & Kalton, G. (2006). On the bias of the multiple-imputation variance estimator in survey sampling. *Journal of the Royal Statistical Society: Series B (Statistical Methodology), 68*, 509-521.

[81] Laird, N. M., & Ware, J. H. (1982). Random-effects models for longitudinal data. *Biometrics, 38*, 963-974.

[82] Lavori, P. W., Dawson, R., & Shera, D. (1995). A multiple imputation strategy for clinical trials with truncation of patient data. *Statistics in Medicine, 14*, 1913-1925.

[83] Lawless, J. F. (1982). *Statistical Models and Methods for Lifetime Data.* New York: John Wiley and Sons.

[84] Li, K. H., Raghunathan, T. E., & Rubin, D. B. (1991a). Large-sample significance levels from multiply imputed data using moment-based statistics and an F-reference distribution. *Journal of the American Statistical Association, 86*, 1065-1073.

[85] Li, K. H., Meng, X. L., Raghunathan, T. E., & Rubin, D. B. (1991b). Significance levels from repeated P-values with multiply-imputed data. *Statistica Sinica, 1*, 65-92.

[86] Lillard, L., Smith, J. P., & Welch, F. (1986). What do we really know about wages? the importance of nonreporting and census imputation. *Journal of Political Economy, 94*(3, Part 1), 489-506.

[87] Ling, W., Amass, L., Shoptaw, S., Annon, J. J., Hillhouse, M., Babcock, D., et al. (2005). A multi-center randomized trial of buprenorphine-naloxone versus clonidine for opioid, detoxification: Findings from the

national institute on drug abuse clinical trials network. *Addiction, 100*, 1090-1100.

[88] Lipsitz, S. R., Parzen, M., & Molenberghs, G. (1998). Obtaining the maximum likelihood estimates in incomplete R × C contingency tables using a poisson generalized linear model. *Journal of Computational and Graphical Statistics, 7*, 356-376.

[89] Little, R. J. A. (1985). Nonresponse adjustments in longitudinal surveys: Models for categorical data. *Bulletin of the International Statistical Institute, Proceedings of the 45th Session: Invited Papers, , Section 15.1.* pp. 1-18.

[90] Little, R. J. A. (2003). The Bayesian approach to sample survey inference. *Analysis of Survey Data* (pp. 49-57). Chichester, West Sussex, England; Hoboken, NJ: John Wiley and Sons.

[91] Little, R. J. A., & Rubin, D. B. (1987). *Statistical Analysis with Missing Data* (1st Edition ed.). New York: John Wiley & Sons.

[92] Little, R. J. A., & Rubin, D. B. (2002). *Statistical Analysis with Missing Data* (2nd Edition ed.). New York: John Wiley & Sons.

[93] Little, R. J. (1993). Statistical analysis of masked data. *Journal of Official Statistics, 9*, 407-426.

[94] Little, R. J. A. (1982). Models for nonresponse in sample surveys. *Journal of the American Statistical Association, 77*, 237-250.

[95] Little, R. J. A. (1992). Regression with missing X's: A review. *Journal of the American Statistical Association, 87*, 1227-1237.

[96] Little, R. J. A. (1994). A class of pattern-mixture models for normal incomplete data. *Biometrika, 81*, 471-483.

[97] Little, R. J. A. (1995). Modeling the drop-out mechanism in repeated-measures studies. *Journal of the American Statistical Association, 90*, 1112-1121.

[98] Little, R. J. A., Liu, F., & Raghunathan, T. E. (2004; 2005). Statistical disclosure techniques based on multiple imputation. *Applied Bayesian*

Modeling and Causal Inference from Incomplete-data perspectives (pp. 141-152). John Wiley & Sons, Ltd.

[99] Little, T. D., & Rhemtulla, M. (2013). Planned missing data designs for developmental researchers. *Child Development Perspectives, 7*, 199-204.

[100] Lumley, T. S. (2010). *Complex Surveys: A guide to analysis using R.* New York: John Wiley & Sons.

[101] Mallinckrodt, C. H., Clark, S. W., Carroll, R. J., & Molenbergh, G. (2003). Assessing response profiles from incomplete longitudinal clinical trial data under regulatory considerations. *Journal of Biopharmaceutical Statistics, 13*, 179-190.

[102] Mallinckrodt, C. H., Clark, W. S., & David, S. R. (2001). Type I error rates from mixed effects model repeated measures versus fixed effects anova with missing values imputed via last observation carried forward. *Drug Information Journal, 35*, 1215-1225.

[103] McKendrick, A. G. (1926). Applications of mathematics to medical problems. *Proceedings of Edinburgh Mathematical Society, 44*, 98-130.

[104] Meng, X. (1994). Multiple-imputation inferences with uncongenial sources of input. *Statistical Science, 9*, 538-558.

[105] Meng, X., & Rubin, D. B. (1992). Performing likelihood ratio tests with multiply-imputed data sets. *Biometrika, 79*, 103-111.

[106] Mislevy, R. J., Beaton, A. E., Kaplan, B., & Sheehan, K. M. (1992). Estimating population characteristics from sparse matrix samples of item responses. *Journal of Educational Measurement, 29*, 133-161.

[107] Molenberghs, G., & Verbeke, G. (2005). *Models for repeated discrete data.* U.S.A.: Springer.

[108] Molenberghs, G., Beunckens, C., Jansen, I., Thijs, H., Verbeke, G., & Kenward, M. (2014). Missing data. In W. Ahrens, & I. - Pigeot (Eds.), *Handbook of epidemiology* (pp. 1283-1335) Springer New York.

[109] National Research Council. (2010). *The Prevention and Treatment of Missing Data in Clinical Trials.* Washington, DC: The National Academies Press.

[110] Neter, J., Kutner, M., Wasserman, W & Nachtsheim, C. (1996). *Applied Linear Statistical Models*. New York: McGraw-Hill/Irwin.

[111] Nordheim, E. V. (1984). Inference from nonrandomly missing categorical data: An example from a genetic study on Turner's syndrome. *Journal of the American Statistical Association, 79*, 772-780.

[112] Orchard, T., & Woodbury, M. A. (1972). Missing information principle: Theroy and applications. *Proceedings of the Sixth Berkeley Symposium on Mathematical Statistics and Probability, 1*. pp. 697-715.

[113] Padilla, M. A., Divers, J., Vaughan, L. K., Allison, D. B., & Tiwari, H. K. (2009). Multiple imputation to correct for measurement error in admixture estimates in genetic structured association testing. *Human Heredity, 68*, 65-72.

[114] Pearson, K. (1910). *Supplement to the memoir entitled: The Influence of Parental Alcoholism on the Physique and Ability of the Offspring*. London: Dulau & Co., LTD.

[115] Pearson, K. (1911). *An attempt to correct some of the misstatements made by Sir Victor Horsely, F.R.S., F.R.C.S., and Mary D. Sturge, M.D., in their criticisms of the Galton Laboratory Memoir: 'A first study of the influence of parental alcoholism, [ill]rc.'*. London: Dulau & Co., LTD.

[116] Peytchev, A. (2012). Multiple imputation for unit nonresponse and measurement error. *Public Opinion Quarterly, 76*, 214-237.

[117] Politz, A., & Simmons, W. (1949). An attempt to get the "not at homes" into the sample without callbacks. *Journal of the American Statistical Association, 44*, 9-16.

[118] Potter, F. (1988). Survey of procedures to control extreme sampling weights. *ASA Proceedings of the Section on Survey Research Methods, American Statistical Association*, pp. 453-458.

[119] Potter, F. (1990). A study of procedures to identify and trim extreme sampling weights. *Proceedings of the Section on Survey Research Methods, American Statistical Society*, pp. 225-230.

[120] Pregibon, D. (1977). Typical survey data: Estimation and imputation. *Survey Methodology, 2,* 70-102.

[121] Preisser, J. S., Lohman, K. K., & Rathouz, P. J. (2002). Performance of weighted estimating equations for longitudinal binary data with dropouts missing at random. *Statistics in Medicine, 21,* 3035-3054.

[122] Raghunathan, T., & Paulin, G. (1998). Multiple imputation of income in the consumer expenditure survey: Evaluation of statistical inference. *Proceedings of the Section on Business and Economic Statistics of the American Statistical Association,* pp. 1-10.

[123] Raghunathan, T. E. (1987). Large sample significance levels from multiply-imputed data. (Doctoral Dissertation, Harvard University).

[124] Raghunathan, T. E. (2006). Combining information from multiple surveys for assessing health disparities. *Allgemeines Statistisches Archiv, 90,* 515-526.

[125] Raghunathan, T. E., Lepkowski, J. M., Hoewyk, J. V., & Solenberger, P. (2001). A multivariate technique for multiply imputing missing values using a sequence of regression models. *Survey Methodology, 27,* 85-95.

[126] Raghunathan, T. E., Reiter, J. P., & Rubin, D. B. (2003). Multiple imputation for statistical disclosure limitation. *Journal of Official Statistics-Stockholm, 19,* 1-16.

[127] Raghunathan, T. E., & Grizzle, J. E. (1995). A split questionnaire survey design. *Journal of the American Statistical Association, 90,* 54-63.

[128] Raghunathan, T. E., & Siscovick, D. S. (1998). Combining exposure information from various sources in an analysis of a case-control study. *Journal of the Royal Statistical Society: Series D (the Statistician), 47,* 333-347.

[129] Rao, J. N. K. (1996). On variance estimation with imputed survey data. *Journal of the American Statistical Association, 91,* 499-506.

[130] Rao, J. N. K., & Shao, J. (1992). Jackknife variance estimation with survey data under hot deck imputation. *Biometrika, 79,* 811-822.

[131] Reiter, J. P., & Raghunathan, T. E. (2007). The multiple adaptations of multiple imputation. *Journal of the American Statistical Association, 102*, 1462-1471.

[132] Reiter, J. P. (2003). Inference for partially synthetic, public use micro-data sets. *Survey Methodology, 29*, 181-188.

[133] Reiter, J. P. (2005). Releasing multiply imputed, synthetic public use microdata: An illustration and empirical study. *Journal of the Royal Statistical Society: Series A (Statistics in Society), 168*, 185-205.

[134] Reiter, J. P. (2007). Small-sample degrees of freedom for multi-component significance tests with multiple imputation for missing data. *Biometrika, 94*, 502-508.

[135] Reiter, J. P., Raghunathan, T. E., & Kinney, S. K. (2006). The importance of modeling the sampling design in multiple imputation for missing data. *Survey Methodology, 32*, 143-149.

[136] Robins, J. M., Rotnitzky, A., & Zhao, L. P. (1994). Estimation of regression coefficients when some regressors are not always observed. *Journal of the American Statistical Association, 89*, 846-866.

[137] Robins, J. M., Rotnitzky, A., & Zhao, L. P. (1995). Analysis of semi-parametric regression models for repeated outcomes in the presence of missing data. *Journal of the American Statistical Association, 90*, 106-121.

[138] Robins, J. M., & Wang, N. (2000). Inference for imputation estimators. *Biometrika, 87*, 113-124.

[139] Royston, P. (2004). Multiple imputation of missing values. *The Stata Journal, 4*, 227-241.

[140] Rubin, D. B. (1974). Estimating causal effects of treatments in randomized and nonrandomized studies. *Journal of Educational Psychology, 66*, 688-701.

[141] Rubin, D. B. (1976). Inference and missing data. *Biometrika, 63*, 581-592.

[142] Rubin, D. B. (1976a). Comparing regressions when some predictor values are missing. *Technometrics, 18*, 201-205.

[143] Rubin, D. B. (1976b). Noniterative least squares estimates, standard errors and F-tests for analyses of variance with missing data. *Journal of the Royal Statistical Society. Series B (Methodological), 38*, 270-274.

[144] Rubin, D. B. (1977). Formalizing subjective notions about the effect of nonrespondents in sample surveys. *Journal of the American Statistical Association, 72*, 538-543.

[145] Rubin, D. B. (1978a). Multiple imputations in sample surveys-a phenomenological Bayesian approach to nonresponse. *Proceedings of the Survey Research Methods Section of the American Statistical Association, 1*, 20-34.

[146] Rubin, D. B. (1978b). Bayesian inference for causal effects: The role of randomization. *The Annals of Statistics, 6*, 34-58.

[147] Rubin, D. B. (1987). *Multiple Imputation for Nonresponse in Surveys* (99th ed.) New York: John Wiley and Sons.

[148] Rubin, D. B. (1993). Discussion statistical disclosure limitation. *Journal of Official Statistics, 9*, 461-468.

[149] Rubin, D. B. (1996). Multiple imputation after 18+ years. *Journal of the American Statistical Association, 91*, 473-489.

[150] Rubin, D. B. (2003). Discussion on multiple imputation. *International Statistical Review, 71*, 619-625.

[151] Rubin, D. B. (2005). Causal inference using potential outcomes. *Journal of the American Statistical Association, 100*, 322-331.

[152] Rubin, D. B., & Schenker, N. (1986). Multiple imputation for interval estimation from simple random samples with ignorable nonresponse. *Journal of the American Statistical Association, 81*, 366-374.

[153] Sarndal, C. E. (2007). The calibration approach in survey theory and practice. *Survey Methodology, 33*, 99-119.

[154] Sarndal, C. E., Swensson, B., & Wretman, J. (2013). *Model Assisted Survey Sampling*. New York: Springer.

[155] Sautory, O. (2003). Proceedings of the statistics Canada symposium 2003. *Calimar 2: A New Version of the Calmar Calibration Adjustment Program.*

[156] Schafer, J. L. (1997). *Analysis of Incomplete Multivariate Data.* New York: CRC Press.

[157] Schenker, N., Raghunathan, T. E., & Bondarenko, I. (2010). Improving on analyses of self-reported data in a large-scale health survey by using information from an examination-based survey. *Statistics in Medicine, 29*, 533-545.

[158] Seaman, S. R., & White, I. R. (2013). Review of inverse probability weighting for dealing with missing data. *Statistical Methods in Medical Research, 22*, 278-295.

[159] Shao, J. (1996). Bootstrap model selection. *Journal of the American Statistical Association, 91*, 655-665.

[160] Shoemaker, D. M. (1973). *Principles and Procedures of Multiple Matrix Sampling.* Cambridge, MA, Ballinger Pub. Co.

[161] Siddiqui, O., & Ali, M. W. (1998). A comparison of the random-effects pattern mixture model with last-observation-carried-forward (LOCF) analysis in longitudinal clinical trials with dropouts. *Journal of Biopharmaceutical Statistics, 8*, 545-563.

[162] Simmons, W. R. (1954). A plan to account for "not-at-homes" by combining weighting and callbacks. *The Journal of Marketing, 19*, 53.

[163] Soldo, B. J., Hurd, M. D., Rodgers, W. L., & Wallace, R. B. (1997). Asset and health dynamics among the oldest old: An overview of the AHEAD study. *The Journals of Gerontology. Series B, Psychological Sciences and Social Sciences, 52 Spec No*, 1-20.

[164] Splawa-Neyman, J., Dabrowska, D. M., & Speed, T. P. (1923 [1990]). On the application of probability theory to agricultural experiments. Essay on principles. section 9. *Statistical Science, 5*, 465-472.

[165] Stasny, E. A. (1986). Estimating gross flows using panel data with nonresponse: An example from the Canadian labour force survey. *Journal of the American Statistical Association, 81*, 42-47.

[166] Stigler, S. M. (1999). *Statistics on the Table*. Cambridge, Massachusetts: Harvard University Press.

[167] Sturge, M., & Horsley, V. (1911). On some of the biological and statistical errors in the work on parental alcoholism by Miss Elderton and Professor Karl Pearson, F.R.S. *British Medical Journal, 1*(2611), 72-82.

[168] Sukhatme, P. V., & Sukhatme, B. V. (1970). *Sampling Theory of Surveys with Applications* (2nd ed.). New Delhi, India: Asia Publishing House.

[169] Thomas, N., Raghunathan, T. E., Schenker, N., Katzoff, M. J., & Johnson, C. L. (2006). An evaluation of matrix sampling methods using data from the national health and nutrition examination survey. *Survey Methodology, 32*, 217.

[170] Thomas, N., & Gan, N. (1997). Generating multiple imputations for matrix sampling data analyzed with item response models. *Journal of Educational and Behavioral Statistics, 22*, 425-445.

[171] Vach, W. (1994). *Logistic Regression with Missing Values in the Covariates*. New York: Springer-Verlag.

[172] Valliant, R., Dever, J. A., & Krueter, F. (2013). *Practical Tools for Designing and Weighting Survey Samples*. New York Heidelberg Dordrecht London: Springer.

[173] van Buuren, S. (2012). *Flexible Imputation of Missing Data*. Boca Raton, FL, Chapman and Hall/CRC.

[174] van Buuren, S., & Oudshoorn, K. (1999). Flexible multivariate imputation by MICE. *Technical Report, Leiden: TNO Preventie En Gezondheid, TNO/VGZ/PG 99.054.*

[175] Verbeke, G., & Molenberghs, G. (2000). *Linear Mixed Models for Longitudinal Data,* New York: Sprinter Verlag.

[176] Von Hippel, P. T. (2007). Regression with missing Y's: An improved strategy for analyzing multiply imputed data. *Sociological Methodology, 37*, 83-117.

[177] Wacholder, S., Carroll, R. J., Pee, D., & Gail, M. H. (1994). The partial questionnaire design for case-control studies. *Statistics in Medicine*, *13*(5-7), 623-634.

[178] Wang, N., & Robins, J. M. (1998). Large-sample theory for parametric multiple imputation procedures. *Biometrika, 85*, 935-948.

[179] Weisberg, S. (2013). *Applied linear regression* (4th ed.). New York, John Wiley and Sons.

[180] Wilks, S. S. (1932). Moments and distributions of estimates of population parameters from fragmentary samples. *Annals of Mathematical Statistics, 3*, 195.

[181] Yates, F. (1933). The analysis of replicated experiments when the field results are incomplete. *The Empire Journal of Experimental Agriculture, 1*, 129-142.

[182] Zeger, L. M., & Thomas, N. (1997). Efficient matrix sampling instruments for correlated latent traits: Examples from the national assessment of educational progress. *Journal of the American Statistical Association, 92*, 416-425.

[183] Zheng, H., & Little, R. J. A. (2003). Penalized spline model-based estimation of the finite populations total from probability-proportional-to-size samples. *Journal of Official Statistics, 19*, 99-107.

[184] Zhou, H. (2014). Accounting for complex sample designs in multiple imputation using the finite population Bayesian bootstrap. (Doctoral Dissertation, University of Michigan, 2014).

Index